Praise for
Concept-Based Ma...

"I attended a Concepts Based Curriculum training course led by Jennifer Wathall and was really inspired by what I learned. Far too often, as teachers, we can become narrowly focused on the topics that we are covering; with concepts there is a whole new opportunity for students to understand the big ideas and the connections between different subjects. Jennifer skillfully guided us through how we can introduce students to a concept-based curriculum. I was really impressed with the method of writing generalizations which provide a framework for exploration. These generalizations can in fact make the focus of a lesson or series of lessons much more exciting, allowing students to break out of the constraints of a limited topic range. Ultimately, I left the course determined to try out a concepts-based model with a new unit we are developing on human rights. With the conceptual lens, this promises to be a much more thought-provoking unit for our students."

—*John Edwards, Head of History Department*
King George V School, Hong Kong

"Secondary teachers are constantly being encouraged to change their practice but few books have addressed the topics of secondary mathematics or given examples that secondary teachers can relate to. This book does that. Another strength is the connection of the content to the math processes and practices—the heart of good instruction. The figures provided to summarize big ideas are excellent. I love the potential of this book for using it as a text for middle and secondary teachers, a guide for professional development, and a place for individual reflection. I know for sure that I would use it for my student teacher seminar class and anytime that I was instructing upper level math teachers. I've been waiting for this!"

—*Barbara Fox, Adjunct Professor, Student Teacher Supervisor*
University of Massachusetts, Lesley University, Regis College

"Jennifer Wathall's *Concept-Based Mathematics* is one of the most forward-thinking mathematics resources on the market. While highlighting the essential tenets of Concept-Based Curriculum design, her accessible explanations and clear examples show how to move students to deeper conceptual understandings. This book ignites the mathematical mind!"

—*Lois A. Lanning, Author*
Designing a Concept-Based Curriculum for English Language Arts, K–12

"One of the major strengths of *Concept-Based Mathematics* is the examples and the visual diagrams that outline major topics. The book provides rubrics that allow teachers to determine where they are in their methodology and an opportunity to decide where they can improve their teaching strategies."

—*Amanda McKee, High School Mathematics Instructor*
Florence County School District #5

"Jennifer Wathall provides a solid rationale, backed up by numerous practical and authentic examples to increase the quality of conceptual math classroom teaching practices so needed to develop the next generation learners. This is a must have for any secondary school's professional library."

—*Dave Nagel, Author Consultant*
Corwin / NZJ Learning (LLC)

"The major strength of *Concept-Based Mathematics* is that it addresses teaching mathematics in a way that invites students to learn and encourages not only content knowledge, but depth of knowledge, rigor, and critical thinking. Inquiry is a means of synergizing your classroom and drawing students in so that they want to learn. This is going to require a change of mind-set for teachers and the administrators alike as well as training. This book provides an opportunity for both."

—*Pamela L. Opel, Intervention Specialist*
Gulfport School District

"Jennifer Wathall's book offers a clear understanding of how complex learning of mathematics is, and how to use this understanding to create a thinking classroom. She explains, from a scientific point of view and in a very well-researched and comprehensive manner, sprinkled with plenty of practical examples, what is the interdependence between the processes and knowledge and how to plan, run, and assess for a concept-based mathematics classroom. This book is a must-read for all mathematics educators, a book that should definitely be on the table in each mathematics office."

—*Dr. Daniela Vasile, Head of Mathematics*
South Island School, Hong Kong

"Wathall is a master at covering all the bases here; this book is bursting with engaging assessment examples, discussion questions, research, and resources that apply specifically to mathematical topics. Any math teacher or coach would be hard-pressed to read it and not come away with scores of ideas, assessments, and lessons that she could use instantly in the classroom. As an IB Workshop Leader and instructional coach, I want this book handy on a nearby shelf for regular referral—it's a boon to any educator who wants to bring math to life for students the world over."

—*Alexis Wiggins, Instructional Coach,*
IB Workshop Leader and Consultant

Concept-Based Mathematics

Concept-Based Mathematics

Teaching for Deep Understanding in Secondary Classrooms

Jennifer T. H. Wathall

Foreword by H. Lynn Erickson

FOR INFORMATION:

Corwin

A SAGE Company

2455 Teller Road

Thousand Oaks, California 91320

(800) 233-9936

www.corwin.com

SAGE Publications Ltd.

1 Oliver's Yard

55 City Road

London EC1Y 1SP

United Kingdom

SAGE Publications India Pvt. Ltd.

B 1/I 1 Mohan Cooperative Industrial Area

Mathura Road, New Delhi 110 044

India

SAGE Publications Asia-Pacific Pte. Ltd.

3 Church Street

#10-04 Samsung Hub

Singapore 049483

Certified Sourcing
www.sfiprogram.org
SFI-00453

Acquisitions Editor: Erin Null

Senior Associate Editor: Desirée A. Bartlett

Editorial Assistant: Andrew Olson

Production Editor: Amy Schroller

Copy Editor: Diana Breti

Typesetter: C&M Digitals (P) Ltd.

Proofreader: Jennifer Grubba

Indexer: Judy Hunt

Cover Designer: Anupama Krishnan

Marketing Manager: Rebecca Eaton

Library of Congress Cataloging-in-Publication Data

Wathall, Jennifer.

Concept-based mathematics : teaching for deep understanding in secondary classrooms / Jennifer Wathall.

pages cm
Includes bibliographical references and index.

ISBN 978-1-5063-1494-5 (pbk. : alk. paper)

1. Mathematics—Study and teaching—Psychological aspects. 2. Mathematics—Study and teaching (Middle school) 3. Mathematics—Study and teaching (Secondary) I. Title.
QA11.2.W376 2016
510.71'2—dc23   2015031139

This book is printed on acid-free paper.

Cover image: *Red Magic II* by Suman Vaze
36" × 20", acrylic on canvas

The 3 × 3 magic square can be constructed in only one way so that the verticals, horizontals, and diagonals all add to 15. This work is another way of looking at the magic square.

16 17 18 19 20 10 9 8 7 6 5 4 3 2 1

Contents

List of Figures   x

Foreword   xiv

    H. Lynn Erickson

Preface   xvi

Acknowledgments   xxi

About the Author   xxiii

Author's Note   xxiv

PART I. WHAT IS CONCEPT-BASED CURRICULUM AND INSTRUCTION IN MATHEMATICS: RESEARCH AND THEORY

1. **Why Is It Important for My Students to Learn Conceptually?**   2

    Why Do We Need to Develop Curriculum and Instruction to Include the Conceptual Level?   2

    The Structure of Knowledge and the Structure of Process   3

    Applying the Structure of Knowledge and the Structure of Process   9

    Teaching for Inquiry   13

    Chapter Summary   23

    Discussion Questions   24

2. **What Are the Levels of the Structures of Knowledge and Process for Mathematics?**   25

    The Levels of the Structure of Knowledge   26

    The Levels of the Structure of Process   30

    Macro, Meso, and Micro Concepts in Mathematics   52

    The Marriage of the Structure of Knowledge and the Structure of Process   53

    Two-Dimensional vs. Three-Dimensional Curriculum Models   56

    Chapter Summary   59

    Discussion Questions   60

PART II. HOW TO CRAFT GENERALIZATIONS AND PLAN UNITS OF WORK TO ENSURE DEEP CONCEPTUAL UNDERSTANDING

3. What Are Generalizations in Mathematics?   62

    What Is the Difference Between a
        Generalization and a Principle in Mathematics?   63

    How Do We Craft Quality Mathematics Generalizations?   69

    How Do We Draw Out Conceptual Understandings From Our Students?   74

    Chapter Summary   82

    Discussion Questions   82

4. How Do I Plan Units of Work for a Concept-Based Curriculum?   83

    Unit Webs   85

    Unit Planning   87

    Guiding Questions   88

    Planning a Unit of Work for Functions   89

    Planning a Unit of Work for Circles   97

    Planning a Unit of Work for Calculus: Differentiation and Integration   97

    Chapter Summary   127

    Discussion Questions   127

PART III. HOW DO WE ENGAGE STUDENTS THROUGH INSTRUCTIONAL PRACTICE? STRATEGIES TO ENGAGE AND ASSESS

5. How Do I Captivate Students? Eight Strategies
   for Engaging the Hearts and Minds of Students   130

    Strategy 1: Create a Social Learning Environment   131

    Strategy 2: Provide an Open, Secure Environment
        to Allow for Mistakes as Part of the Learning Process   133

    Strategy 3: Use Appropriate Levels of Inquiry and
        Employ Inductive Approaches to Develop Conceptual Understanding   134

    Strategy 4: Reduce Whole Class Teacher Talk Time   135

    Strategy 5: Cater to Everyone in your Class; Use Differentiation Strategies 141

    Strategy 6: Assessment Strategies   149

    Strategy 7: Be Purposeful When Asking Students
        to Answer Questions; There is Safety in Numbers   151

    Strategy 8: Flexible Fronts: Arranging your Classroom   151

    Chapter Summary   154

    Discussion Questions   155

6. How Do I Know My Students Understand the Concepts? Assessment Strategies   156

    Assessments With Conceptual Depth   157

    Open Inquiry Tasks and Open-Ended Questions   158

Visible Thinking Routines		160
Performance Assessment Tasks		164
The Frayer Model		172
Concept Attainment Cards		172
Agree, Disagree, and Depends		172
Zero, One, Two, or Three		176
Assessing and Developing Core Transdisciplinary Skills		177
Assessing the Developing Concept-Based Student		178
Self-Assessments		180
Chapter Summary		185
Discussion Questions		186

7. How Do I Integrate Technology to Foster Conceptual Understanding? 187

Mathematics Graphing Software and Graphical Display Calculators		191
Flipped Classroom		192
Multimedia Projects		199
Collaboration Tools: Google Applications		202
Apps on Mobile Devices		203
When Not to Use Technology		205
Chapter Summary		214
Discussion Questions		214

8. What Do Ideal Concept-Based Mathematics Classrooms Look Like? 215

Foster a Culture of Growth Mindset		216
Pedagogical Principles in an Ideal Classroom		217
Developing the Ideal Concept-Based Mathematics Lesson		218
Developing Concept-Based Lesson Planning in the Ideal Mathematics Classroom		221
Common Concerns and Misconceptions About Concept-Based Curriculum and Instruction		228
Chapter Summary		233
Last Words		233
Discussion Questions		235

Glossary	237
Teaching for Deep Understanding in Secondary Schools Book Study	243
References and Further Reading	255
Index	260

List of Figures

Chapter 1

Figure 1.1 Side by Side: The Structure
of Knowledge and the Structure of Process   5

Figure 1.2 The Structure of Knowledge for Functions   6

Figure 1.3 The Structure of Process for Functions   8

Figure 1.4 The Structure of Knowledge
and the Structure of Process for Functions, Side by Side   10

Figure 1.5 Inductive vs. Deductive Approaches   12

Figure 1.6 Two-Dimensional vs. Three-Dimensional
Curriculum/Instruction Models   13

Figure 1.7 Developing Intellect Through
Inquiry Process Continuum Model   14

Figure 1.8 Levels of Inquiry   16

Figure 1.9 Levels of Inquiry Hierarchy   16

Figure 1.10 Levels of Inquiry for Proving the Pythagorean Theorem Task   17

Figure 1.11 A Structured Inquiry Example   18

Figure 1.12 A Guided Inquiry Example   19

Figure 1.13 An Open Inquiry Example   20

Chapter 2

Figure 2.1 The Structure of Knowledge for Trigonometry   26

Figure 2.2 The Fundamental Theorem
of Calculus in the Structure of Knowledge   29

Figure 2.3 The Structure of Process for Trigonometry   31

Figure 2.4 Comparison of Mathematics
Processes in Different Parts of the World   33

Figure 2.5 The Six Mathematical Processes,
Algorithms, Strategies, and Skills   35

Figure 2.6 An Example of a PEMDAS Algorithm   36

Figure 2.7 Examples of Six Mathematical Processes
Broken Down Into Skills and Strategies   37

Figure 2.8 An Example of Reasoning and Proof   39

Figure 2.9 Number Talks Template   42

Figure 2.10 Example of a Student's Number Talk   43

Figure 2.11 An Example of Making Connections and Creating
        Representations: Graphic Organizer for the Number System   46

Figure 2.12 The Different Uses of the Grid Method   50

Figure 2.13 Examples of Macro, Meso, and Micro Concepts in Mathematics   53

Figure 2.14 Examples of Micro Concepts in
        the International Baccalaureate Diploma Mathematics Courses   54

Figure 2.15 How Processes, Skills, and Facts
        Contribute to an Understanding of Concepts   55

Figure 2.16 Structures Example Side by
        Side for the Meso Concept Trigonometry   57

Figure 2.17 Table of Values   58

Chapter 3

Figure 3.1 The Two Types of Enduring
        Understandings: Overarching and Topical   64

Figure 3.2 The Structure of Process for Equations   65

Figure 3.3 The Structure of Knowledge for Vectors   66

Figure 3.4 The Structure of Process for Logarithms   67

Figure 3.5 The Structure of Knowledge for Quadratics   68

Figure 3.6 Side by Side: The Structure of Knowledge
        and the Structure of Process for Quadratics   70

Figure 3.7 Scaffolding Template   71

Figure 3.8 Scaffolding Generalization for Sequences and Series   73

Figure 3.9 Scaffolding a Process Generalization   73

Figure 3.10 An Example of Inductive Inquiry to Draw a Generalization   75

Figure 3.11 An Example of a Graphic Organizer to
        Draw Generalizations From Students for Trigonometry   80

Figure 3.12 Strategies to Draw Generalizations From Students   81

Chapter 4

Figure 4.1 Unit Web for Functions   84

Figure 4.2 Examples of Conceptual Lenses for Mathematics   85

Figure 4.3 Generic Mathematics Unit Web Template   86

Figure 4.4 Step-by-Step Unit Planning Checklist   87

Figure 4.5 Examples of Debatable/Provocative Questions   89

Figure 4.6 Unit Planner for Functions   90

Figure 4.7 Weekly Planner for Functions   94

Figure 4.8 Teacher Notes for Coordinates Game   96

Figure 4.9 Unit Web for Circle Geometry   98

Figure 4.10 Unit Planner for Circle Geometry   99

Figure 4.11 Meso Concept: Calculus Unit Web   103

Figure 4.12 Unit Planner for Calculus   104

Figure 4.13 Calculus Weekly Planner   109

Figure 4.14 Gradients and Slopes   111

Figure 4.15 Increasing and Decreasing Functions   115

Figure 4.16 Stationary Points   116

Figure 4.17 The Product Rule   117

Figure 4.18 Real-Life Problems for Calculus   120

Figure 4.19 Student Solutions to Real-Life Problems for Calculus   123

Figure 4.20 Integration   126

Chapter 5

Figure 5.1 Placemat Activity   132

Figure 5.2 An Example of a Structured Inquiry Task   136

Figure 5.3 An Example of a Guided Inquiry Task   141

Figure 5.4 Student's Response to a Structured Inquiry Task   142

Figure 5.5 Tri-Mind Activity on Functions   145

Figure 5.6 Student's Example of an Analytical Approach to Functions   146

Figure 5.7 Hint Cards for Functions   148

Figure 5.8 Know, Want, Learn Routine   149

Figure 5.9 The Learning Curve: When You Learn Something New…   150

Figure 5.10 No Front! An Example of How to Structure Your Classroom   153

Chapter 6

Figure 6.1 Examples of Inquiry-Based
Assessment Questions and Prompts   159

Figure 6.2 Examples of Visible Thinking Routines   161

Figure 6.3 More Examples of Visible Thinking Routines   162

Figure 6.4 Connect, Extend, Challenge Thinking Routine   163

Figure 6.5 Mathematics Performance Task on Linear Functions   166

Figure 6.6 Using the RAFTS Model to Design a Performance Task   166

Figure 6.7 Performance Task: Setting a Scenario   167

Figure 6.8 The Trigonometric Ratios of Similar Right-Angled Triangles   168

Figure 6.9 How to Use Trigonometry to Measure the Heights of Buildings   170

Figure 6.10 The Frayer Model Template   173

Figure 6.11 Example of the Frayer Model for Polygons   174

Figure 6.12 Table of Examples and Non-Examples for Rational Numbers   175

Figure 6.13 Probe for Agree, Disagree, or Depends   176

Figure 6.14 Assessing Approaches to Learning Skills   179

Figure 6.15 Rubric: The Developing Concept-Based Student   181

Figure 6.16 Example of a Self-Assessment Worksheet   183

Chapter 7

Figure 7.1 SAMR Integration of Technology Model   189

Figure 7.2 The TPACK Model Framework   190

Figure 7.3 Using Graphing Software   193

Figure 7.4 Using Graphing Software for Circle Theorems   196

Figure 7.5 Flipped Classroom Lesson on Complex Numbers   200

Figure 7.6 Sine Curves Using Spaghetti   206

Figure 7.7 The Unit Circle   209

Figure 7.8 Triangle Inequality   213

Chapter 8

Figure 8.1 Rubric: Concept-Based Instruction by Dr. Lois A. Lanning   222

Figure 8.2 Mastery Level for Concept- Based Lesson Planning   224

Figure 8.3 The Developing Concept-Based
           Teacher: Concept-Based Lesson Planning   225

Book Study Resources

A Participant's Metacognition Log   244

Foreword

H. Lynn Erickson

How many times have you heard the lament, "I was so bad at mathematics in school?" Yet, those with an affinity for mathematics view it as a beautiful abstract language that cuts across fields of knowledge to solve problems, raise questions, explain mysteries, and create wondrous works of art. Jennifer Wathall is one of those people with this affinity. She desires to share her understanding and passion for mathematics with the world. How lucky we are!

I wish all of my prior mathematics teachers had been able to read Jennifer's book and learn from her. As I reflect on my years as a student and my mathematics education specifically, I remember feeling confused as we drilled on daily computations and struggled with word problems. I could *do* math but I did not *understand* math. In my own journey as a teacher, I came to realize the critical importance of conceptual understanding across all of the disciplines. Why had I not been trained to teach for deeper conceptual understanding?

Secondary mathematics teachers across the world will appreciate reading Jennifer's insights about the other half of the equation—the conceptual understanding of mathematics. In traditional mathematics education, we have "assumed" students understand the concepts of mathematics if they could perform the algorithms. It was a step forward as we required students to "explain their thinking" on mathematical problems, but this still did not ensure that students really understood the conceptual relationships inherent in the problem. Jennifer shows us that students need to demonstrate and verbalize their conceptual understanding of mathematics as well as apply it across multiple contexts.

Concept-Based Mathematics: Teaching for Deep Understanding in Secondary Classrooms is a clear explanation of the content and process structures of mathematics supported by salient examples. Jennifer provides practical, engaging, and meaningful learning experiences that draw students to the beauty and power of mathematical concepts and their relationships.

One of the strengths of concept-based curriculum and instruction models is that they are not "programs." They are the explicit expression of the previously implied design principles for quality curriculum design and pedagogy. They overlay *any* curriculum and instruction model and should not be a choice. They are the principles that ensure deeper conceptual thinking and the transfer of understandings through time, across cultures, and across situations. Whether school curricula are textbook based or teacher developed, they must reflect the three-dimensional design principles expressed in this book and other books on concept-based curriculum and instruction, or they will remain a lower level, two-dimensional design model—coverage of facts and skills.

This book is cutting edge. It is the next step to bring mathematics education into the 21st century. It needs to be in the hand of every secondary mathematics teacher and teacher educator. All students deserve to experience the wonder and passion for mathematics that Jennifer so obviously feels. It is time to change the age-old lament to empowering testimonials—"I love mathematics!" "I can use mathematics every day to think and create!" "Math is fun!" "I get math!"

Preface

Purpose of the Book

Traditional curriculum focuses on rules and procedures with little understanding of the conceptual relationships of mathematics—and mathematics is a *language* of conceptual relationships. Traditional curriculum *assumes* the deep understanding of concepts and fails to teach for transferability or to consider context. This book expands and develops the work of Lynn Erickson and Lois Lanning on concept-based curriculum into the realm of mathematics.

> This book expands and develops the work of Lynn Erickson and Lois Lanning on concept-based curriculum into the realm of mathematics.

When learning math, students need to be given time and space to explore and discover the beauty and creativity in math without being fearful of mistakes. Math anxiety exists because of an overemphasis on the processes and skills of this discipline. This book addresses how to create concept-based and inquiry-led curriculum and instruction with a goal to make math enjoyable and accessible to all of our students.

Concept-based curriculum is a three-dimensional design model of curriculum and instruction that frames factual content and skills and processes with disciplinary concepts, generalizations, and principles. In concept-based curriculum and instruction, the development of intellect is achieved through higher order, synergistic thinking in which teachers use the facts, processes, and skills in concert with the concepts, generalizations, and principles. A traditional two-dimensional design model for curriculum and instruction focuses on factual content, processes, and skills and assumes conceptual understanding. The research and consensus on the benefits of developing conceptual understanding is undeniable. Concept-based curriculum produces deeper emotional and intellectual engagement in learning and therefore develops attributes such as critical thinking, reasoning, and creativity.

The intention of this book is to extend the work of Lynn Erickson and Lois Lanning on the Structure of Knowledge and the Structure of Process specifically to mathematics and to help math educators understand how to convey mathematical concepts and ideas using the vehicle of inquiry. All definitions used in the Structure of Knowledge and the Structure of Process derive from the work of Lynn Erickson and Lois Lanning. We need to help students understand that everyone is capable doing math and it is not a matter of whether you can or can't do math.

Special Features

Special features include sample lessons, samples of student work, vignettes from international educators, and discussion questions that may be used in a book study with fellow teachers or in a professional development setting. As an individual teacher or as part of a study group, read each chapter and use the discussion questions at the end of each chapter to reflect on your own practice. Metacognition logs are included at the end of the book, to help you to process, synthesize, and self-reflect on each chapter of the book. There is a chapter on integrating technology to enhance learning and conceptual understanding (Chapter 7) and a Glossary to guide you through the terms used in the book.

The main text is accompanied by a suite of free online resources, which include more sample instructional units and templates for worksheets that foster deeper conceptual understanding of particular math topics for secondary school.

After reading this book, you will be able to focus instruction on deeper conceptual understandings and equip students for future success. It will provide you with practical examples of concept-based lessons, unit webs, unit planners, and different assessment tools to enable you to develop a concept-based approach to your curriculum and instruction.

Concept-based mathematics is grounded in the philosophy that in order to develop intellect, instruction and curriculum needs to focus on the big ideas or conceptual understandings. This can be achieved through instructional practice and designing tasks that do not sacrifice the content or rigor of any prescribed syllabus. In fact, concept-based curriculum challenges students to employ higher order thinking skills. Concept-based curriculum can overlay any curriculum, such as the Common Core State Standards and Basal curriculum (United States), GCSE and A Levels (UK), as well as the International Baccalaureate Middle Years (MYP) and Diploma (DP) mathematics programs.

> In this ever changing, dynamic and complex world, mathematics education must engage students intellectually and emotionally.

In this ever changing, dynamic and complex world, mathematics education must engage students intellectually and emotionally. The ability to think conceptually, transfer understandings across contexts and situations, and to enjoy learning and problem solving are major goals for mathematics education today so we can prepare our students for future success. Technological advancements of even the last decade have influenced instruction, and the key to utilizing technology effectively is not *what* tool is being used but *how* the technology is used to enhance learning.

I hope this book inspires you on your journey to develop conceptual understanding in your students and to eradicate math anxiety and fear by fostering a growth mindset. I hope you will join me on this journey for this much-needed math education reform.

How to Use the Companion Website

http://www.resources.corwin.com/WathallConceptBasedMathematics

The companion website offers the following resources to supplement this book:

- Straightforward activities designed to help teachers understand and apply concept-based curriculum and instruction;
- Examples that model each aspect of concept-based curricula;
- Blank templates for designing unit planners and writing quality generalizations;
- Guiding questions to help you and your book study group to reflect on the process of implementation and next steps;
- A metacognition log: a powerful tool for self-reflection that focuses on the end-of-chapter discussion questions in this book.

If you are working with a book study group or PLC, you might want to upload the activities, templates, companion website discussion questions, and the metacognition log to a cloud on an app such as Google Drive so that you can share your personal written reflections with your team as you write them.

Here are some suggestions for how an individual educator can use the website:

- Read *Concept-Based Mathematics: Teaching for Deep Understanding in Secondary Schools,* ensuring you address the discussion questions at the end of each chapter, and write a reflection on the metacognition log template. You can write your reflections on a piece of paper, on your personal computer, or you can upload them to a cloud to share with your book study group.
- Make notes on areas you would like to develop from the main book.
- Go through the website, using the templates to create your own examples, and answer the discussion questions.
- Think about a unit of work you would like to develop to ensure more conceptual understanding and use the templates to support your planning.
- Trial your ideas in the classroom and modify accordingly.
- Share with colleagues.

Here are some suggestions for how a group of educators can use the website as a book study. Read one chapter a week and meet with the book study group to discuss them.

- Read a chapter of *Concept-Based Mathematics: Teaching for Deep Understanding in Secondary Schools* and write a reflection in the metacognition log.
- Share your metacognition log notes, one chapter at a time, with your book study group.
- Use the discussion questions from each chapter to stimulate sharing of ideas during your meetings with your study group.
- Go through the website, using the templates to create your own examples.
- Think about a unit of work you would like to develop to ensure more conceptual understanding, and use the templates to support your planning. This could be a collaborative effort with three or four other teachers.
- Trial your ideas in the classroom.
- Share and review your unit planners with colleagues, either in person or in a cloud.

Audience

Drawn from my 24 years as an international educator and presenter, this book will uncover the secrets to help all students in middle and high school understand how to convey the conceptual language of mathematics. This book is intended for middle and high school teachers, trainee teachers in undergraduate education programs, and graduate education courses ranging from bachelor of education, diploma in education, to masters in arts specializing in education.

Chapter Overview

Part I (Chapters 1 and 2) of this book discusses what a concept-based curriculum looks like for mathematics and explains, in detail, Lynn Erickson's Structure of Knowledge and Lois Lanning's Structure of Process applied to the topic of functions. Examples of levels of inquiry (structured and guided) and inductive teaching are given. The key to inductive teaching is that students draw and form generalizations by working on specific examples initially.

Part II (Chapters 3 and 4) guides readers in the practice of applying concept-based curriculum and instruction to math. Chapter 3 deals with crafting generalizations, which are statements of conceptual understanding. Lynn Erickson provides a three-step guide to writing quality generalizations, which are statements of what we want our students to understand from their program of study. Chapter 4 includes models of unit webs and unit planners.

Part III (Chapters 5 through 8) looks at instructional strategies to intellectually and emotionally engage students to ensure deep conceptual understanding. Chapter 5 discusses eight strategies for lesson planning and captivating your students' hearts and minds. Chapter 6 looks into more detail about formative assessment strategies to track student learning. Chapter 7 discusses how to integrate technology effectively and gives practical activities and digital tools that support conceptual understanding. These tools include using mathematical graphing software, flipped classroom models, multimedia projects, collaborative digital tools, and various educational apps for the classroom. Chapter 8 looks at the elements of an ideal math classroom. It includes rubrics to support the developing concept-based teacher and for developing concept-based instruction. Chapter 8 also addresses common concerns and misconceptions about concept-based curriculum and instruction.

After reading this book,

- You will have a better understanding of the benefits of a concept-based instructional design model;
- You will be able to overlay a concept-based curriculum and instruction model onto any curriculum and implement it in your classroom; and
- You will have ideas and resources to engage your students and increase their conceptual understanding and enjoyment of mathematics.

Acknowledgments

There are very few people you meet who have such an impact on your life as H. Lynn Erickson has had on mine. I feel so fortunate that Lynn has chosen me to mentor and guide me through my journey into concept-based mathematics. I will forever be indebted for the care, time, and dedication she has shown me during the writing of this book. Lynn: You are a remarkable educator and you have been a wonderful role model for me, helping me believe that anything is possible. Thank you for showing me that I am only at the beginning of my journey in education.

I wish to also thank the following people:

Lois Lanning for giving me a different perspective, challenging me to think, and for her friendship;

Erin Null for believing in this project and giving me invaluable advice;

Desirée Bartlett for her patience, support, and hard work;

My mother, Mei, who has always been a pillar of strength for me;

My sons, Jordan and Jacob, who constantly surprise me with their achievements and who make me feel so proud to be their mother every day;

My husband, Ken, who was extremely patient when he mostly got one-word answers to any conversations during the writing of this book. Without his love and support I would not be the person I am today.

Additional Acknowledgements

Thank you to Marjut Mäenpää for sharing the idea of the grid method and its application to more challenging problems.

Thank you to Dr. Eileen Dietrich for giving me the sine curve spaghetti idea and Rowdy Boeyink for improving on this worksheet.

Thank you Dr. Daniela Vasile for sharing your wonderful ideas on mathematics teaching with me.

Thank you Suman Vaze for sharing your beautiful artwork with the world and me. You should be very proud of everything you have achieved in your life.

Thank you Paul Chillingworth and Isaac Youssef for giving me valuable reviewer feedback.

Publisher's Acknowledgments

Corwin gratefully acknowledges the contributions of the following reviewers:

Paul Chillingworth
Mathematics Advisor, Maths in
Education and Industry
Trowbridge, Wiltshire, UK

Jason Cushner
School Reform Organizer
Eagle Rock School and Rowland
Foundation
Estes Park, CO

Barbara Fox
Adjunct Professor, Student Teacher
Supervisor
University of Massachusetts
Boston, MA

David Horton
Assistant Superintendent
Hemet Unified School District
Hemet, CA

Amanda McKee
High School Math Instructor
Florence School District #5
Johnsonville, SC

Dave Nagel
Corwin Author Consultant
Zionsville, IN

Pamela L. Opel
Teacher, Intervention Specialist
Gulfport School District
Gulfport, MS

Jamalee Stone
Associate Professor of Mathematics
Education
Black Hills State University
Spearfish, SD

Morris White
High School Math Teacher
Alamosa High School
Alamosa, CO

Isaac Youssef
Higher Level Senior Moderator
International Baccalaureate
Auckland, New Zealand

About the Author

Jennifer T. H. Wathall has been a teacher of mathematics for more than 20 years. She graduated from the University of Sydney with a BSc majoring in mathematics and completed postgraduate studies at the University of Hong Kong.

She has worked in several international schools, including South Island School, Hong Kong; The United Nations International School (UNIS), New York; and she is currently working at Island School, Hong Kong as head of mathematics. In the international arena, she has presented workshops such as "How to Effectively Integrate Multimedia into the Classroom" at the 21st Century Conference in Hong Kong and Shanghai and the Asian Technology Conference in Mathematics, Bangkok and Beijing. She has also given talks around Asia about how to effectively integrate a 1:1 program into the mathematics classroom.

As a qualified International Baccalaureate workshop leader ("Mathematics, Concepts, and Inquiry in the Diploma Program and Approaches to Teaching and Learning"), Jennifer has delivered numerous workshops in the Asia Pacific region. Her role as a field representative for the IB Asia Pacific serves as part of the quality assurance framework. She has consulted for IB mathematics textbooks and has developed an IB Category 3 workshop on "The Use of the Casio GDC in IB Mathematics." Jennifer has delivered presentations at the IB Asia Pacific Conference ("Using Inquiry in the IB Mathematics Classroom") and at the IB Americas Annual Conference ("Concept-Based Mathematics"). Currently she is part of the external curriculum review group for IB diploma mathematics based in The Hague and Cardiff. As an expert in IB mathematics, Jennifer serves as an honorary faculty advisor and part-time instructor for the University of Hong Kong.

She is a certified trainer in the DISC™ behavior assessment tool, and she is a certified independent consultant in Concept-Based Curriculum Design by Dr. H. Lynn Erickson. Jennifer works as a consultant helping math departments and schools transition to concept-based curriculum and instruction. She utilizes her skills as a certified performance coach to facilitate transition and change.

Author's Note

I was born to be a teacher. I love being in the classroom and just spending time with my students with the goal of inspiring a love for learning. Nothing excites me more than seeing those light bulb moments during a lesson when students have a gleam in their eye because they get it. That gleam tells me my students understand on a deeper level than what a textbook or video can explain. I have been so lucky that all of my life I knew what my vocation would be.

My father fostered my love of learning and teaching, as he was a teacher himself. He taught English in the air force before joining the diplomatic corps. He read to me most nights: sometimes Jane Austen or Charles Dickens and sometimes famous Chinese fables to teach me about Chinese culture and history. He was patient, intelligent, and possessed a lifelong thirst for knowledge. His passing in July 2014 inspired me to complete my two years of research and to write this book. Math education needs to change. Too many students have been scarred for life because of their negative experiences in math classrooms. Everybody can do math in an environment focused on conceptual understanding and a growth mindset. I hope to start a revolution in math classrooms and help teachers to think and reflect about what they are teaching. What do we want our students to learn and understand, and what is mathematics? Is it a discipline of processes?

Mathematics comes from the Greek word *máthēma,* which means "that which is learnt." In Modern Greek, *máthēma* means "to learn." Math lessons need to focus on *learning* and not on *performing.* Many mathematicians have different interpretations of what mathematics is. Below are my favorite quotes from mathematicians.

Pure mathematics is, in its way, the poetry of logical ideas.

—Albert Einstein, German-born theoretical physicist
and 1921 Nobel Prize winner, 1879–1955

Nature's great book is written in mathematics.

—Galileo Galilei; Italian physicist, mathematician,
astronomer, and philosopher; 1564–1642

Mathematics is the queen of sciences and number theory is the queen of mathematics. She often condescends to render service to astronomy and other natural sciences, but in all relations she is entitled to the first rank.

—Carl Friedrich Gauss; German
mathematician, physicist, and prodigy; 1777–1855

A mathematician, like a painter or poet, is a maker of patterns. If his patterns are more permanent than theirs, it is because they are made with ideas.

—Godfrey H. Hardy, English
mathematician known for his achievements
in number theory and mathematical analysis, 1877–1947

Mathematics is a more powerful instrument of knowledge than any other that has been bequeathed to us by human agency.

—René Descartes; French philosopher,
mathematician, scientist, and writer; 1596–1650

The essence of mathematics is not to make simple things complicated, but to make complicated things simple.

—Stan Gudder, mathematics professor, University of Denver

Whenever I am in a social situation and tell someone I am a math teacher, I receive one of two reactions: anxiety alongside an alarming panic, with people expressing how much they hated math at school; and the less common response—how much they loved math—which begins a lively conversation about the usefulness of math. The first response saddens me. How can mathematics elicit such fear and negativity? English, art, and even science teachers do not elicit such strong emotions in people. A longstanding tradition sees mathematics as an elusive discipline that few could comprehend. Many people recall negative experiences when learning mathematics that have instilled fear of the discipline. Timed tasks, rote memorization of formulae with little conceptual understanding, and a focus on performance have created math fear and reinforce these negative experiences.

As a person who made mathematics education her career, I fortunately did not have those negative experiences as a child. I loved the challenge of puzzles and problems that were presented and possessed a passion for mathematics throughout my school life. When I was 12 years old, my mother took me to a fortune teller in Taiwan who looked into my eyes and said I would follow my passion to become a math teacher. Who knows if the fortune teller could really tell, but from that day as a child, I felt I knew my destiny and have been fortunate enough to be able to share my joy for math education well into my third decade.

In loving memory of my father,
David Kuo Cheng Chang,
who inspired me to be a lifelong learner

1929–2014

Part I

What Is Concept-Based Curriculum and Instruction in Mathematics?

Research and Theory

....

Why Is It Important for My Students to Learn Conceptually?

Around the world, mathematics is highly valued and great importance is placed on learning mathematics. Private tutors in non–Asian countries serve a remedial purpose, whereas in Asia, everyone has a tutor for providing an increased knowledge base and skill development practice. Many students in Asia enroll in programs like "Kumon," which focus on practicing skills (which has its place) and "doing" math rather than "doing and understanding" math. When you ask students who are well rehearsed in skills to problem solve and apply their understanding to different contexts, they struggle. The relationship between the facts, skills, and conceptual understandings is one that needs to be developed if we want our students to be able to apply their skills and knowledge to different contexts and to utilize higher order thinking.

Why Do We Need to Develop Curriculum and Instruction to Include the Conceptual Level?

According to Daniel Pink (2005), author of *A Whole New Mind,* we now live in the Conceptual Age. It is unlike the Agricultural Age, Information Age, or the Industrial Age because we no longer rely on the specialist content knowledge of any particular person. The Conceptual Age requires individuals to be able to critically think, problem solve, and adapt to new environments by utilizing transferability of ideas. "And now we're progressing yet again—to a society of creators and empathizers, of pattern recognizers and meaning makers" (Pink, 2005, p. 50).

Gao and Bao (2012) conducted a study of 256 college-level calculus students. Their findings show that students who were enrolled in concept-based learning environments scored higher than students enrolled in traditional learning environments. Students in the concept-based learning courses also liked the approaches more. A better grasp of concepts results in increased understanding and transferability.

With the exponential growth of information and the digital revolution, success in this modern age requires efficient processing of new information and a higher level of abstraction. Frey and Osborne (2013) report that in the next two decades, 47% of jobs in the United States will no longer exist due to automation and computerization. The conclusion is that we do not know what new jobs may be created in the next two decades. Did cloud service specialists, android developers, or even social marketing companies exist 10 years ago?

How will we prepare our students for the future? How will our students be able to stand out? What do employers want from their employees? It is no longer about having a wider knowledge base in any one area.

Hart Research Associates (2013) report the top skills that employers seek are the following:

- Critical thinking and problem solving,
- Collaboration (the ability to work in a team),
- Communication (oral and written), and
- The ability to adapt to a changing environment.

How do we develop curriculum and instruction to prepare our students for the future?

We owe our students more than asking them to memorize hundreds of procedures. Allowing them the joy of discovering and using mathematics for themselves, at whichever level they are able, is surely a more engaging, interesting and mind-expanding way of learning. Those "A-ha" moments that you see on their faces; that's why we are teachers.

David Sanda, Head of Mathematics
Chinese International School, Hong Kong

The Structure of Knowledge and the Structure of Process

Knowledge has a structure like other systems in the natural and constructed world. Structures allow us to classify and organize information. In a report titled *Foundations for Success,* the U.S. National Mathematics Advisory Panel (2008) discussed three facets of mathematical learning: the factual, the procedural, and the conceptual.

These facets are illustrated in the Structure of Knowledge and the Structure of Process, developed by Lynn Erickson (2008) and Lois Lanning (2013).

The **Structure of Knowledge** is a graphical representation of the relationship between the topics and facts, the concepts that are drawn from the content under study, and the generalization and principles that express conceptual relationships (transferable understandings). The top level in the structure is Theory.

Theory describes a system of conceptual ideas that explain a practice or phenomenon. Examples include the Big Bang theory and Darwin's theory of evolution.

The **Structure of Process** is the complement to the Structure of Knowledge. It is a graphical representation of the relationship between the processes, strategies, skills and concepts, generalizations, and principles in process–driven disciplines like English language arts, the visual and performing arts, and world languages.

For all disciplines, there is interplay between the Structure of Knowledge and the Structure of Process, with particular disciplines tipping the balance beam toward one side or another, depending on the purpose of the instructional unit. The Structure of Knowledge and the Structure of Process are complementary models. Content-based disciplines such as science and history are more knowledge based, so the major topics are supported by facts. Process-driven disciplines such as visual and performing arts, music, and world languages rely on the skills and strategies of that discipline. For example, in language and literature, processes could include the writing process, reading process, or oral communication, which help to understand the author's craft, reader's craft, or the listener's craft. These process-driven understandings help us access and analyze text concepts or ideas.

 Both structures have concepts, principles, and generalizations, which are positioned above the facts, topics, or skills and strategies. Figure 1.1 illustrates both structures. Figure 1.1 can also be found on the companion website, to print out and use as a reference.

The Structure of Knowledge and the Structure of Process for Functions

Mathematics *can* be taught from a purely content-driven perspective. For example, functions can be taught just by looking at the facts and content; however, this does not support learners to have complete conceptual understanding. There are also processes in mathematics that need to be practiced and developed that could also reinforce the conceptual understandings. Ideally it is a marriage of the two, which promotes deeper conceptual understanding. Figure 1.2 illustrates the Structure of Knowledge for the topic of functions.

Topics organize a set of facts related to specific people, places, situations, or things. Unlike history, for example, mathematics is an inherently conceptual language, so "Topics" in the Structure of Knowledge are actually *broader concepts,* which break down into micro-concepts at the next level.

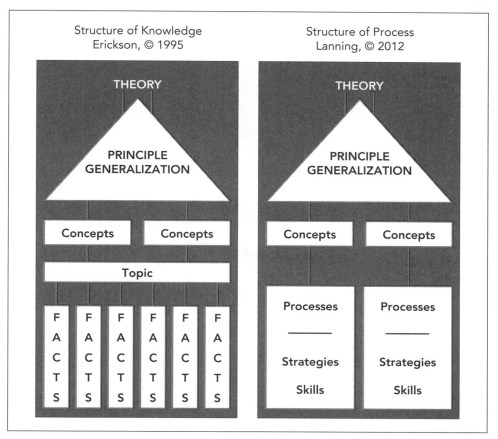

© 2014 H. Lynn Erickson and Lois A. Lanning
Transitioning to Concept-Based Curriculum and Instruction, Corwin Press Publishers, Thousand Oaks, CA.

As explained by Lynn Erickson (2007), "The reason mathematics is structured differently from history is that mathematics is an inherently conceptual language of concepts, subconcepts, and their relationships. Number, pattern, measurement, statistics, and so on are the broadest conceptual organizers" (p. 30).

More about concepts in mathematics will be discussed in Chapter 2.

Facts are specific examples of people, places, situations, or things. Facts do not transfer and are locked in time, place, or a situation. In the functions example seen in Figure 1.2, the facts are $y = \mathrm{m}x + \mathrm{c}$, $y = \mathrm{a}x^2 + \mathrm{b}x + \mathrm{c}$, and so on. The factual content in mathematics refers to the memorization of definitions, vocabulary, or formulae. When my student knows the *fact* that $y = \mathrm{m}x + \mathrm{c}$, this does not mean she understands the *concepts* of linear relationship, y–intercept, and gradient.

According to Daniel Willingham (2010), automatic factual retrieval is crucial when solving complex mathematical problems because they have simpler problems

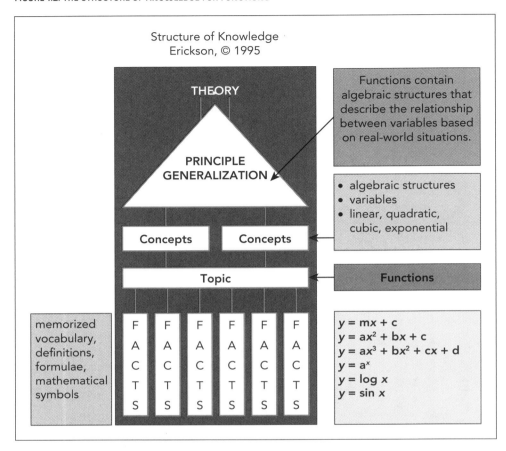

Structure of Knowledge
Erickson, © 1995

THEORY

PRINCIPLE
GENERALIZATION

Functions contain algebraic structures that describe the relationship between variables based on real-world situations.

Concepts   Concepts

Topic

- algebraic structures
- variables
- linear, quadratic, cubic, exponential

Functions

memorized vocabulary, definitions, formulae, mathematical symbols

FACTS FACTS FACTS FACTS FACTS FACTS

$y = mx + c$
$y = ax^2 + bx + c$
$y = ax^3 + bx^2 + cx + d$
$y = a^x$
$y = \log x$
$y = \sin x$

Adapted from original Structure of Knowledge figure from *Transitioning to Concept-Based Curriculum and Instruction*, Corwin Press Publishers, Thousand Oaks, CA.

embedded in them. Facts are the critical content we wish our students to know, but they do not themselves provide evidence of deep conceptual understanding.

Formulae, in the form of symbolic mathematical facts, support the understanding of functions. This leads to a more focused understanding of the *concepts* of linear functions, quadratic functions, cubic functions, exponential functions, variables, and algebraic structures in Figure 1.2. The generalization "Functions contain algebraic structures that describe the relationship between two variables based on real-world situations" is our ultimate goal for conceptual understanding related to the broad concept of functions. Please take a look at the companion website for more examples of the Structure of Knowledge and the Structure of Process on the topic of linear functions. See Figures M1.1 and M1.2.

Concepts are mental constructs, which are timeless, universal, and transferable across time or situations. Concepts may be broad and abstract or more conceptually specific to a discipline. "Functions" is a broader concept, and the micro-concepts at the next level are algebraic structures, variables, linear, quadratic, cubic, and exponential. Above the concepts in Figure 1.2 are the principles and generalizations.

Principles and **generalizations** are transferable understandings that allow students to make connections between two or more concepts. In mathematics, the principles are the theorems, the cornerstone truths. Though generalizations and principles are both statements of conceptual relationship, the principles do not contain a qualifier such as *often, can,* or *may* because they are immutable "truths" as we know them. Because generalizations do not rise to the level of a law or theorem, they *may* require a qualifier if they do not hold true in all cases. Principles and generalizations are often exemplified in a real-life context for mathematics; however, they are not exclusively portrayed in this way. In Figure 1.2, another generalization could have been the following: "Algebraic tools allow highly complex problems to be solved and displayed in a way that provides a powerful image of change over time" (Fuson, Kalchman, & Bransford, 2005, p. 351).

> Generalizations are statements that connect two or more concepts.

Although the Structure of Knowledge provides the deep understanding of the content of mathematics, the processes, strategies, and skills also provide important conceptual understanding.

The Structure of Process represents the procedural facet of learning mathematics. Processes, skills, and strategies are included in the lowest levels in the Structure of Process. "Skills are smaller operations or actions that are embedded in strategies, and when appropriately applied 'allow' the strategies to work. Skills underpin a more complex strategy" (Lanning, 2013, p. 19).

Strategies are systematic plans that learners consciously adapt and monitor to improve learning performance. As explained by Erickson and Lanning (2014), "Strategies are complex because many skills are situated within a strategy. In order to effectively employ a strategy, one must have control over a variety of the skills that support the strategy." (p. 46). An example of a strategy in math would be making predictions or drawing conclusions.

Processes are actions that produce results. A process is continuous and moves through stages during which inputs (materials, information, people's advice, time, etc.) may transform or change the way a process flows. A process defines what is to be done—for example, the writing process, the reading process, the digestive process, the respiratory process, and so on.

Figure 1.3 illustrates an example of the mathematical process of creating representations and the generalizations associated with this mathematical process. Throughout this functions unit, students will learn different strategies and skills that support the process of creating representations. This could include using a table of values or an algebraic or geometric form of a function.

Concepts that can be drawn from this process include substitution, revision, interpretation, and models. Two or more of the concepts are used to write unit generalizations, which are also known as process generalizations. The process generalizations in Figure 1.3 are as follows:

Mathematicians create different representations—table of values, algebraic, geometrical—to compare and analyze equivalent functions.

....

The revision of a mathematical model or substitution of data may enhance or distort an accurate interpretation of a problem.

....

When students are guided to these generalizations, they demonstrate their understanding of the creating representations process.

FIGURE 1.3: STRUCTURE OF PROCESS EXAMPLE FOR FUNCTIONS

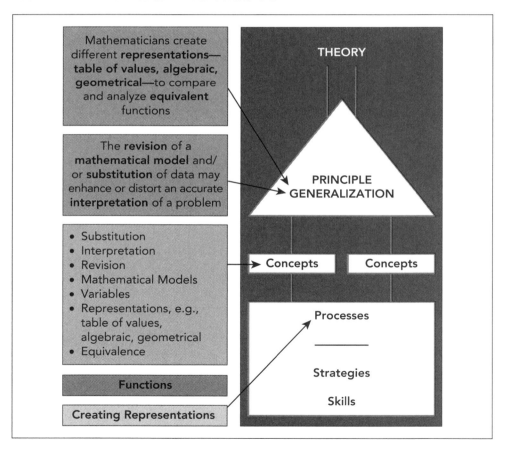

Adapted from original Structure of Process figure from *Transitioning to Concept-Based Curriculum and Instruction*, Corwin Press Publishers, Thousand Oaks, CA.

Other strategies and skills, such as graphing and analytical skills, support the process of creating representations. This process supports the concepts of mathematical models, substitution, interpretation, revision, variables, equivalence, and so on.

In Figure 1.4, we look at the dual part that the Structure of Knowledge and the Structure of Process each play in ensuring a deep understanding of content and process in mathematics. For the concept of functions, we include the content that needs to be learned as well as the skills and strategies that are employed fluently to aid the process of creating representations. The ability to employ strategies and skills fluently is referred to as *procedural fluency*. Visit the companion website to see additional summaries of the components of the Structures of Knowledge and Process. See Figures M1.3 and M1.4.

To help understand the generalization "Functions contain algebraic structures that describe the relationship between two variables based on real-world situations," we work to ensure the conceptual relationships are revealed. The concepts of algebraic structures, variables, linear, quadratic, and cubic help us connect the facts to give mathematical content more meaning and promote deeper understanding. The mathematical process involved is creating representations, and it supports the understanding of the concepts substitution, interpretation, revision, variables, mathematical models, and equivalence. Mathematical processes will be discussed in detail in the next chapter.

The language of mathematics is different to languages like English and Chinese. There are things that are strictly allowed and there are things that are strictly not. It is the formal nature of the language that often causes confusion and errors in learners. However, over-emphasis on the formality, and some teachers are only concerned with practicing formal exercises, prevents understanding of the beauty, creativity, and utility of mathematics.

Chris Binge, Principal Island School, Hong Kong

Applying the Structure of Knowledge and the Structure of Process

Inductive vs. Deductive Teaching

In my first years of teaching, it was common practice in the mathematics classroom to adopt the PPP model (presentation, practice, and production) of **deductive, teacher-led instruction**. The PPP approach typically looks like this:

Step 1: Teacher introduces the formula, such as the Pythagorean theorem, and demonstrates three working examples.

Step 2: Ask students to practice using the formula.

Step 3: Ask students to produce their own examples.

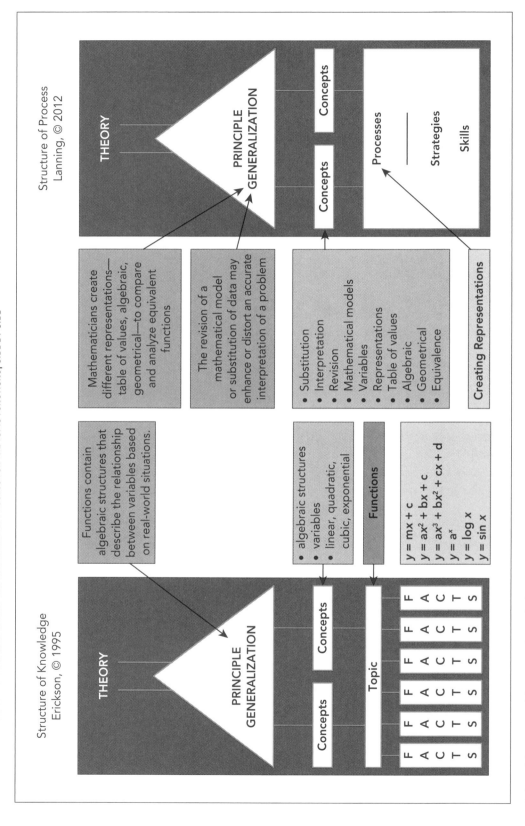

Adapted from original Structure of Knowledge and Structure of Process figures from *Transitioning to Concept-Based Curriculum and Instruction*, Corwin Press Publishers, Thousand Oaks, CA.

The **two-dimensional model of instruction**, which focuses on the facts and content of the subject and the rote memorization of procedures and topics, is intellectually shallow. A two-dimensional curriculum and instruction model focuses on the bottom levels of the Structure of Knowledge and the Structure of Process. This encourages students to work at a low-order level of thinking (such as memorization of facts or perfunctory performance of lower level skills) in a content/skill-based, coverage-centered curriculum. A two-dimensional model often presents the generalization or new concepts at the beginning of the learning cycle and follows a direct teaching methodology.

This is typical of a deductive approach in teaching. I have witnessed many, many lessons utilizing this approach, and to me, this is like telling our students what the present is before they open it! The concept-based model is generally an inductive teaching model that draws the understandings from the students as a result of structured or guided inquiry.

An **inductive approach**, like mathematical induction, allows learners to start with specific examples and form generalizations for themselves. In his research on how the brain learns mathematics, David Sousa (2015) states that the

> Inductive approaches lead to generalization formation.

human brain is a powerful pattern seeker, and we have an innate number sense or what scientists call "numerosity." The inductive approach utilizes this innate quality for number sense and pattern finding. The teacher acts as a facilitator, helping students to discover relationships and seek patterns for themselves.

The **three-dimensional model of instruction** suggests a more sophisticated design with a third level: the conceptual level. In a three-dimensional curriculum and instruction model, the lower levels of the Structure of Knowledge and the Structure of Process are important components, but the third dimension of concepts, principles, and generalizations ensures that conceptual thinking and understanding are prominent.

A three-dimensional, inductive approach encourages students to construct generalizations at the end of the learning cycle through the use of inquiry. As stated by Erickson and Lanning (2014), "Deep understanding and the transfer of knowledge and skills require that teachers understand the relationship between the factual/skill level and the conceptual level, and use this relationship effectively in instruction" (p. 23).

Figure 1.5 illustrates the difference between inductive and deductive approaches.

FIGURE 1.5: INDUCTIVE VS. DEDUCTIVE APPROACHES

Deductive Approach	Students are given the generalizations at the beginning of a lesson	Students then practice the generalizations through specific examples
Inductive Approach	Students are given specific examples at the beginning of the lesson	Students construct generalizations from

An inductive model is a student-centered approach, helping students to think logically and scientifically and allowing students to generalize by utilizing higher order thinking. Discovering inductive approaches changed my entire teaching practice and influences every student learning experience I plan for my students. The inductive approach provides a framework; it is a structure for all mathematical concepts to be conveyed to students in an analytical, coherent fashion. The key to inductive teaching is that students draw and form generalizations by working on specific examples initially.

Introducing the Pythagorean theorem utilizing an inductive approach would look like this:

1. Look at the following right-angled triangles and work out the squares of each of the sides. (Students work out specific numerical examples.)
2. What generalization can you make about the relationship between all three sides when they are squared? (Students now generalize by pattern seeking.)

Bransford, Brown, and Cocking (2000) offer a comprehensive survey of neurological and psychological research that provides strong support for constructivism and inductive methods. "All new learning involves transfer of information based on previous learning" (p. 53).

Inductive instruction presents new information in the context of situations, issues, and problems to which students can relate, so there is a much greater chance that the information can be linked to their existing cognitive structures. John D. Bransford et al. (2000) explain, "Motivation to learn affects the amount of time students are willing to devote to learning. Learners are more motivated when they can see the usefulness of what they are learning and when they can use it to do something that has an impact on others" (p. 61).

Inductive methods, such as problem-based learning, support techniques that use authentic situations and problems.

Generalizations and principles in the Structure of Knowledge and the Structure of Process are timeless, universal, transcend cultures, and are transferable ideas. They allow the learner to connect the facts and concepts for deeper meaning and understanding. The three-dimensional model of curriculum and instruction, according to Erickson and Lanning (2014), includes concepts, generalizations, and principles to ensure that curriculum and instruction focus on intellectual depth, the transfer of understanding, and the development of conceptual brain schemata. The three-dimensional model is contrasted with the traditional two-dimensional model of coverage and memorization.

Figure 1.6 illustrates the two-dimensional model, also known as the "inch deep, mile wide" approach to curriculum. In contrast, the three-dimensional model represents a more comprehensive, sophisticated design for curriculum and instruction.

FIGURE 1.6: TWO-DIMENSIONAL VS. THREE-DIMENSIONAL CURRICULUM/INSTRUCTION MODELS

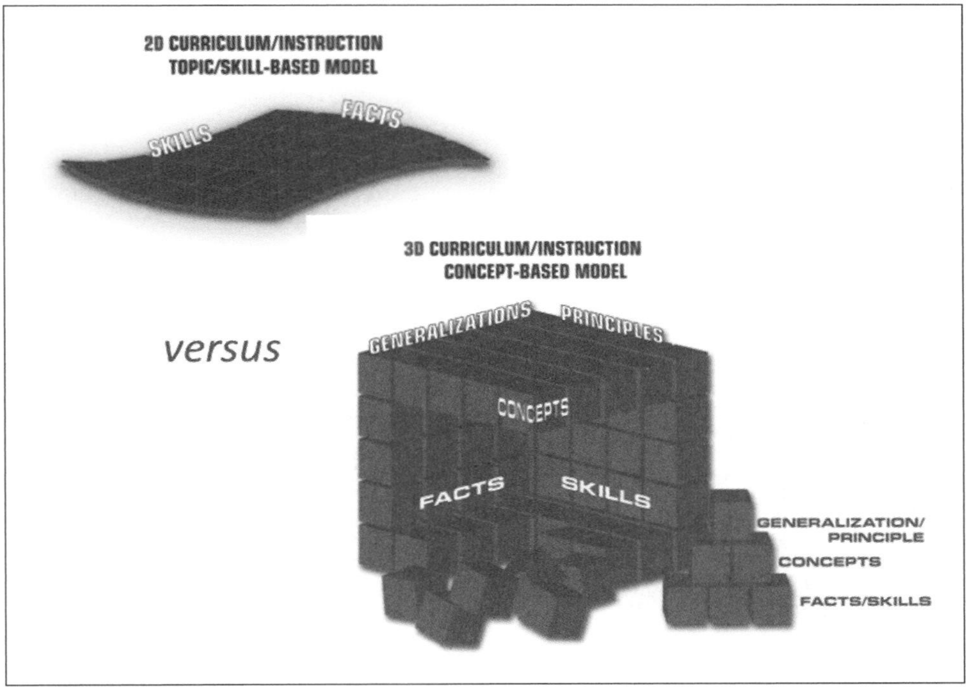

Transitioning to Concept-Based Curriculum and Instruction, Corwin Press Publishers, Thousand Oaks, CA.

Teaching for Inquiry

Inquiry is a vehicle and is about not telling students what the surprise is before opening the present. I have met many teachers in my travels, and often I hear the following about inquiry:

"I don't have time for inquiry! I need to get through the content!"

"I have inquiry lessons once per week!"

"Inquiry just doesn't work with my students; they need to be spoon fed!"

"Inquiry does not work for my students; they do not have the ability!"

Inquiry refers to posing questions, problems, or scenarios rather than providing established facts or knowledge. Inquiry means to seek truth, information, or knowledge, and individuals carry out the natural process of inquiry throughout their lives. Unfortunately, traditional curriculum discourages inquiry; students learn not to ask questions and to accept facts that are given. A study by Gelman, Gruber, and Ranganath (2014) found that learning is more effective when students are curious. Memory is also enhanced when students are in a state of curiosity. Inquiry encourages curiosity in students by posing questions to engage thought and interest.

Through inquiry and a variety of pedagogical approaches, such as cooperative and problem-based learning, students can develop skills for success while understanding the concepts involved (Barron & Darling-Hammond, 2008). Lynn Erickson encapsulates this idea as follows: "Information without intellect is meaningless." Figure 1.7 illustrates the synergistic relationship between the facts, skills, and concepts all being achieved through a continuum of inquiry.

In order to develop intellect in our students we need to establish synergistic thinking through the inquiry continuum.

FIGURE 1.7: DEVELOPING INTELLECT THROUGH INQUIRY PROCESS CONTINUUM MODEL

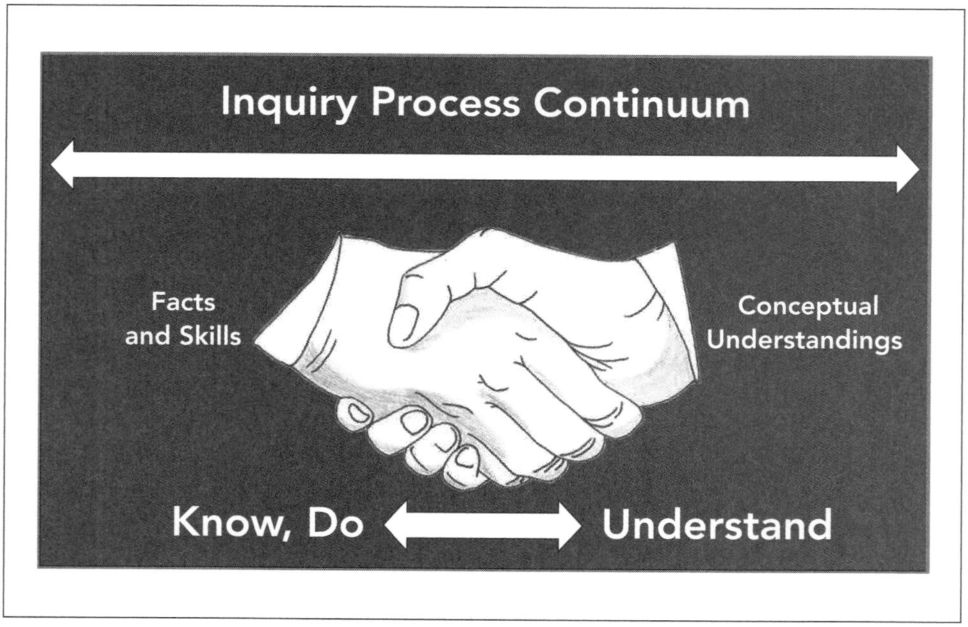

© 2016 Jennifer Wathall

Erickson and Lanning (2014) state that "Synergistic thinking requires the inter-action of factual knowledge and concepts. Synergistic thinking requires a deeper level of mental processing and leads to an increased understanding of the facts related to concepts, supports personal meaning making, and increases motivation for learning" (p. 36).

The vehicle of inquiry is used to foster synergistic thinking. The design of guiding questions in the form of factual, conceptual, and debatable questions also supports synergistic thinking and allows students to bridge the gap between the facts and skills and conceptual understandings.

For additional resources, visit the companion website where you will find an example of a traditional activity as well as guidance on how to facilitate synergis-tic thinking and a template to plan a synergistic student activity of your own. See Figures M1.6 & M1.7.

As an example, in order to understand the concepts of linear functions, parameters, and variables, one must know facts, such as $y = mx + c$ or $Ax + By + C = 0$, and be able to plot points and create different representations. The inquiry process would ask students to investigate linear functions for different values for the parameters m and c. This supports the understanding of the concepts of linear, parameters, variables, and functions. Inquiry also stimulates student motivation and interest and leads to a deeper understanding of transferable concepts.

I have had the pleasure of working with Mike Ollerton, a pioneer in inquiry-based learning of mathematics from the United Kingdom. In his short piece on "Enquiry-Based Learning" (2013) he writes, "The underpinning pedagogy of enquiry-based learning (EBL) is for learners to gain and to use & apply knowledge in ways which places responsibility for the learning upon students. This is at the heart of support-ing independent learning and requires the teacher become a facilitator of students' knowledge construction; as a key aspect of sense making."

Different levels of inquiry are used as appropriate to the context and classroom situation. Figure 1.8 describes the levels of inquiry, adapted from the work of Andrew Blair (http://www.inquirymaths.com). Figure 1.9 shows the hierarchy of the levels of inquiry. The triangle represents the progression of inquiry levels, which can start off being quite narrow and structured, then move to a guided approach, and then ultimately to open inquiry, giving all students more opportu-nities to explore.

The three levels of inquiry—structured, guided, and open—originated in the learning approaches of science-based disciplines (Banchi & Bell, 2008). The important questions here are why and when do we use the different levels of inquiry?

Level	Description
Structured	• Heavily scaffolded • Predictable line of inquiry • Predictable outcomes
Guided	• Different lines of inquiry • Predictable outcomes
Open	• Different lines of inquiry • Unpredictable outcomes

FIGURE 1.9: LEVELS OF INQUIRY HIERARCHY

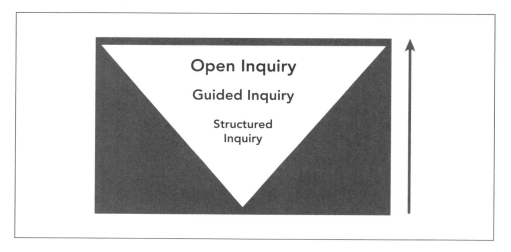

Structured inquiry is heavily scaffolded and suitable perhaps for learners and teachers who are new to inquiry. Structured inquiry fosters confidence in learners while promoting autonomy and independence. Teachers who are not accustomed to using inquiry find it difficult to "let go" of control, and structured inquiry provides a happy medium. The outcomes are predictable and predetermined by the design of the task.

Guided inquiry presents learners with opportunities for different lines of inquiry, with predicable outcomes. For example, ask students for different methods to prove a particular theorem (e.g., the Pythagorean theorem). Guided inquiry has fewer prompts and gives the learner more freedom to choose his or her own pathways to the desired outcome.

Open inquiry promotes different lines of inquiry with unpredictable outcomes. Truly authentic, open inquiry engages the learner's interest and creativity. For example, the International Baccalaureate Mathematics Standard and Higher Levels include an internal assessment called a "personal exploration." Students are asked to choose an area of mathematics, conduct their own research, and draw their own conclusions. One of my past students, who was a ranked Hong Kong tennis champion, chose to write about tennis and binomial theorem. Another student with scoliosis looked at the curvature of her spine over the years using statistical analysis.

Open inquiry is not to be confused with pure "discovery" learning, when very little guidance is given to the learner. There is a misconception that inquiry is about giving students an open problem and letting them "run with it" with little guidance or input from the teacher. This is far from the intention of inquiry. Inquiry is student centered, inherently inductive, and peaks students' motivation and interest. Inquiry is not an excuse for passivity. The teacher's role is vital in facilitating and guiding the students during different stages of learning.

On the following pages there are three examples of student tasks on the same topic: proving the Pythagorean theorem. The topic is presented in three different ways to illustrate structured, guided, and open levels of inquiry.

> **Open inquiry is student centered, with extensive input from the teacher.**

Figure 1.10 summarizes the main features and the difference between the three levels of inquiry for the Pythagorean theorem task. Figures 1.11, 1.12, and 1.13 are the student tasks.

FIGURE 1.10: LEVELS OF INQUIRY FOR PROVING THE PYTHAGOREAN THEOREM TASK

Task: Proving the Pythagorean Theorem	Features
Structured approach	Step-by-step scaffolded questions and prompts The table allows students to calculate the areas of different shapes within the large square and prompts students to find a relationship.
Guided approach	Fewer scaffolded prompts Given the large square with the tilted square inside, students must work out that finding the areas of the shapes inside.
Open approach	Students are asked to research their own proof with hundreds to choose from. They need to explain and show understanding of their proof.

FIGURE 1.11: A STRUCTURED INQUIRY EXAMPLE

Proving Pythagorean Theorem

Find the area of the following shapes and complete this table.

A Area of the large square	B Area of the tilted square	C Area of the four triangles	B + C	Connecting A, B, and C

Explain in words the relationship you have discovered. Use a diagram to illustrate your explanation.

 For a completed version of Figure 1.11, please visit the companion website.

FIGURE 1.12: A GUIDED INQUIRY EXAMPLE

Proving the Pythagorean Theorem

Investigate the relationship between a, b, and c using the following diagram.

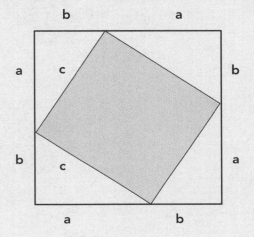

FIGURE 1.13: AN OPEN INQUIRY EXAMPLE

Proving the Pythagorean Theorem

There are hundreds of proofs for the Pythagorean theorem. Research one proof and explain the proof with diagrams. Use any medium to explain your proof. This could include a poster, movie, applet, or Google presentation.

> **T**o state a theorem and then to show examples of it is literally to teach backwards.
>
> E. Kim Nebeuts From Inspirational Quotes,
> Word, Sayings (2015)

Through inductive inquiry, students are given opportunities to find generalizations and patterns they observe from specific examples. Studies have shown that a concept-based curriculum using an inductive approach results in a higher level of retention and conceptual understanding of the content.

> **D**eductive approaches are the norm in traditional math classrooms—we rote-learn processes in a mechanical way without understanding the true reasoning and meaning behind the problem itself. Inquiry-based learning requires us to think and analyze for ourselves, then come up with a conclusion or generalization, which is the fun and beauty behind learning mathematics. We are encouraged to challenge ourselves and step away from our comfort zones in order to expand our knowledge of mathematics. Both learning methods are effective in the short term for an exam. But I have found inductive, inquiry-based approaches allow new information and working methods to be stored in my long-term memory as I actually understand what I am doing.
>
> Chun Yu Yiu, Grade 12 student
> Island School, Hong Kong

According to Borovik and Gardiner (2007, pp. 3–4), the following are some of the top traits of mathematically able students:

- Ability to make and use generalizations—often quite quickly. One of the basic abilities, easily detectable even at the level of primary school: after solving a single example from a series, a child immediately knows how to solve all examples of the same kind.
- Ability to utilize analogies and make connections.
- Lack of fear of "being lost" and having to struggle to find one's way through the problem.

Notice these abilities are described as traits that are not genetic predispositions but qualities that can be nurtured and developed in students. Opportunities to fail or "get stuck" give students the ability to lack fear of being lost or "stuck." In her 2008 Harvard commencement address, J. K. Rowling, author of the *Harry Potter* books, said, "It is impossible to live without failing at something, unless you live so cautiously that you might as well not have lived at all—in which case, you fail by default."

There are three principles outlined in the report *How Students Learn: Mathematics in the Classroom* (Bransford et al., 2005) that are consistent with the concept-based curriculum model:

> Principle 1: Teachers must engage students' preconceptions. (p. 219)

This refers to recognition of students' prior knowledge and prior strategies and the need to build on them to create new strategies and new learning.

> Principle 2: Understanding requires factual knowledge and conceptual frameworks. (p. 231)

This principle suggests the importance of the factual and conceptual and providing a framework for learners to connect the two in the form of generalizations. Learners need to have procedural fluency as well as know the conceptual relationships in order to develop mathematical proficiency.

> Principle 3: A metacognitive approach enables student self-monitoring. (p. 236)

Learners need to be given time and space to explore mathematical concepts—in other words, to self-monitor. More opportunities to reflect on their experiences will help learners to construct their ideas into larger categories and take control of their own learning.

With this overwhelming evidence, you may now ask, how do we develop curriculum and instruction using a concept-based and inquiry-led model? In Chapter 2, we will look at the facts, skills, and strategies in mathematics and how to use them to build conceptual understanding through the Structure of Knowledge and the Structure of Process. Subsequent chapters provide practical activities to guide your journey in developing a three-dimensional concept-based model for curriculum and instruction.

Northside ISD (San Antonio, TX) has been involved in concept-based curriculum for 10 years. It was important for this district that serves 103,000 students to have a K–12 curriculum in all major content areas that was developed using the tenets of concept-based curriculum. Our curriculum staff have been trained and certified by Lynn Erickson. Our teachers and administrators are clear about what our students are expected to know, understand, and do. Concept-based curriculum is without a doubt one of the main reasons Northside ISD continues to be a high performing district.

Linda Mora, Deputy Superintendent for Curriculum
and Instruction Northside ISD,
San Antonio, Texas

Chapter Summary

This chapter laid the foundation for why we need to move from a two-dimensional to a three-dimensional curriculum and instruction model to include the conceptual level. Evidence supports the effectiveness of a concept-based curriculum, which is grounded in an inductive and inquiry-led approach. Concept-based models lead to increased mathematical proficiency and understanding. The chapter discussed what a concept-based curriculum looks like for math and the benefits to students' learning. An overview of the symbiotic relationship between the Structure of Knowledge and the Structure of Process in the realm of mathematics was also provided. Developing intellect requires synergistic thinking, which, according to Lynn Erickson (2007), is an interplay between the factual and conceptual levels of thinking. Synergistic thinking is at the heart of a concept-based curriculum and instruction.

An inductive model is a student-centered approach, helping students to think logically and scientifically, allowing students to generalize by utilizing higher order thinking. The inductive approach provides a framework; it is a structure for all mathematical concepts to be conveyed to students in an analytical, coherent fashion. The key to inductive teaching is that students draw and form generalizations by working on specific examples initially.

Levels of inquiry provide teachers and learners with the opportunity to gain confidence when exploring mathematical concepts. Structured and guided inquiry facilitates differentiation and promotes student and teacher confidence.

Extensive studies in mathematics education indicate a need for curriculum and instruction to include the conceptual level for enduring, deeper understandings. If we are to prepare our students for an unknown future, due to vast technological advances, we must ensure we foster higher order thinking skills.

The next chapter will explain, in detail, the Structure of Knowledge and the Structure of Process as applied to the facts, skills, strategies, and processes of mathematics.

1. Does math education need to undergo a reform? Why or why not?

2. Why do educators need to include the conceptual understandings of a topic represented in a three-dimensional curriculum model?

3. How do the Structures of Knowledge and Process apply to the mathematics realm?

4. What are the features of inductive teaching and the benefits of an inductive approach when learning mathematics?

5. How does synergistic thinking develop intellect?

6. How would you use the different levels of inquiry in your classroom? Think of examples of when you might use each (structured, guided, open).

....

CHAPTER 2

What Are the Levels of the Structures of Knowledge and Process for Mathematics?

Quite a few years ago, in my first lesson on trigonometry with a new class, I asked my students whether they had learned about trigonometric ratios in right-angled triangles. They all replied, "No." When I wrote *SOHCAHTOA* on the board, they said, "Oh, that's what you mean." They all knew the formula $cosx = \dfrac{adjacent}{hypotenuse}$ but did not understand that SOHCAHTOA represents similarity in a set of right-angled triangles sharing a common acute angle. This was a mere memorized fact and algorithm for them, and they had little understanding of the concepts of ratio, similarity, and angles in right-angled triangles. They had been taught to focus on the facts and algorithms first, and there was an assumption of conceptual understanding. These lower levels of thinking are represented in the Structure of Knowledge and the Structure of Process.

As mentioned in Chapter 1, knowledge has an inherent structure, just as the animal and plant kingdoms have structures. With this structure, we are able to classify and recognize similarities, differences, and relationships. Concept-based curriculum requires an understanding of the different levels in the Structure of Knowledge and the Structure of Process and how they affect curriculum design and instruction. Concept-based models include the higher level of intellectual thinking: the conceptual level. An understanding of the Structure of Knowledge and the Structure of Process gives us the ability to plan curriculum and instruction for intellectual development. Let us recap the levels in the Structure of Knowledge.

The Levels of the Structure of Knowledge

The Factual Level

The lowest level in the Structure of Knowledge is the factual level. Factual knowledge includes rote memorization and does not guarantee conceptual depth of understanding.

Facts are specific examples of people, places, situations, or things. They are locked in time, place, or situation. Facts are not transferable and include definitions, formulae in the form of symbols (e.g., $y = mx + c$), and the different names of polygons (e.g., pentagon, hexagon).

Figure 2.1 illustrates the Structure of Knowledge applied to the topic of trigonometry.

FIGURE 2.1: THE STRUCTURE OF KNOWLEDGE FOR TRIGONOMETRY

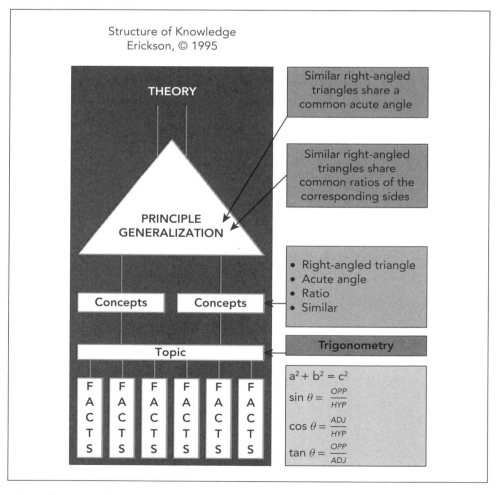

Adapted from original Structure of Knowledge figure from *Transitioning to Concept-Based Curriculum and Instruction*, Corwin Press Publishers, Thousand Oaks, CA.

The factual level includes knowing the shape of the graph of $y = \cos x$ without necessarily understanding how this is generated and memorized formula such as $\cos x = \dfrac{adjacent}{hypotenuse}$. These facts help students to

support the understanding of the concepts of ratio, magnitude, angle, and direction when learning about right-angled trigonometry in a concept-based curriculum.

Knowing a definition or a formula does not imply understanding. Memorized formulae in mathematics are facts that support the broader concepts in mathematics. These facts include the vocabulary, definitions, and formulae in the form of mathematical symbols. For mathematical proficiency and understanding, learners need to know the facts to reinforce their understanding of the related concepts. To *know* means to memorize facts or definitions that are critical to understanding the generalizations (statements of conceptual relationships) for a particular unit.

To continue with the theme of right-angled trigonometry, let us look at the example of the Pythagorean theorem to illustrate this point:

For right-angled triangles, the area of the square drawn from the hypotenuse represents the sum of the areas of the squares drawn from the other sides.

This is a statement of conceptual understanding, which connects the concepts of hypotenuse, area, squares, and sum applied to right-angled triangles. There are numerous inquiry tasks that guide students to understand this principle, one being from http://nrich.maths.org/2293 called "Tilted Squares." In this task, students are asked to spot patterns, make generalizations, and even discover the Pythagorean theorem by finding the areas of tilted squares.

The formula for the Pythagorean theorem is $a^2 + b^2 = c^2$. This is a memorized fact, which does not reflect conceptual understanding. This fact only applies to a specific question, such as the following:

Find c when a = 3 and b = 4.

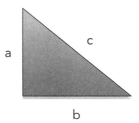

The recall of these facts is highly compressible in the brain and is crucial when problem solving and learning math. Once you understand the process or concept, the brain has an incredible capacity to file this information away for later use—in other words, compress it in the brain.

William Thurston (1990), a Fields Medal winner in mathematics, defined compression particularly well when he wrote, "Mathematics is amazingly compressible: you may struggle a long time, step by step, to work through some process or idea from several approaches. But once you really understand it and have the mental perspective to see it as a whole, there is often a tremendous mental compression. You can file it away, recall it quickly and completely when you need it, and use it as just one step in some other mental process. The insight that goes with this compression is one of the real joys of mathematics (p. 847).

The Difference Between Formulae and Theorems in the Structure of Knowledge

A **formula** is an equation that uses mathematical symbols or variables to show a relationship and is represented by the *facts* in the Structure of Knowledge.

Theorems are statements that have been proven and connect explanations of conceptual understandings. Theorems are represented by *principles* in the Structure of Knowledge.

Let us look at the fundamental theorem of calculus. The first part of this theorem describes the relationship between differentiation and integration as inverse processes of each other. The second part of the fundamental theorem of calculus helps students to evaluate a definite integral without having to go back to the definition of taking the limit of a sum of rectangles.

The fundamental theorem of calculus may also be expressed as a fact or formula in mathematical symbols:

Let $f(x)$ be continuous in the interval [a, b] and $F'(x) = f(x)$, then

$$F(x) = \int_a^x f(t)dt$$

If students know this fact or formula, do they have a deep understanding of the fundamental theorem of calculus? Do they understand the concept of integration as being an inverse process of differentiation or understand that calculus allows the evaluation of a definite integral without having to go back and take the limiting sum of a bounded area?

Figure 2.2 shows the fundamental theorem of calculus depicted in the Structure of Knowledge.

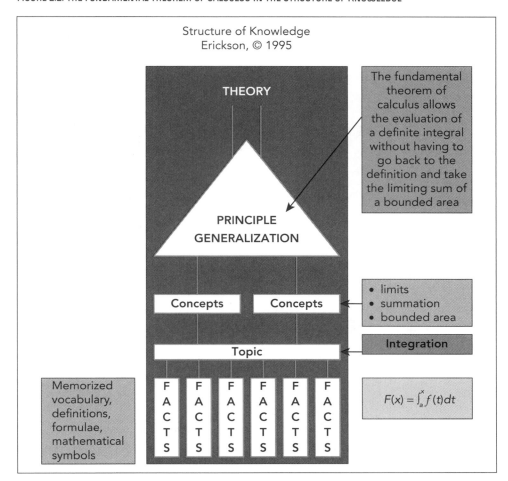

Adapted from original Structure of Knowledge figure from *Transitioning to Concept-Based Curriculum and Instruction,* Corwin Press Publishers, Thousand Oaks, CA.

The Topic and Concepts Levels

Topics in math are broader concepts that break down to specific micro concepts at the next level. In Figure 2.2, the topic "integration" is supported by the micro concepts "limits," "summation," and "bounded areas." More on the classification of math concepts will be discussed later in this chapter.

The Generalizations and Principles Level

Generalizations and principles are statements of conceptual understanding that allow students to make connections between two or more concepts. In mathematics, the principles are theorems—the cornerstone truths. In Figure 2.1, the generalizations are as follows:

Similar right-angled triangles share a common acute angle

> **S**imilar right-angled triangles share common ratios of corresponding sides

In Figure 2.2, the fundamental theorem of calculus represents a principle:

> **T**heorems in calculus allow the evaluation of a definite integral without having to go back to the definition and take the limiting sum of a bounded area. .

Crafting quality generalizations requires an investment of time. How to craft generalizations will be discussed in the next chapter.

The Levels of the Structure of Process

Mathematical Processes, Algorithms, Strategies, and Skills in the Structure of Process

Students in my first trigonometry class were able to easily find lengths of sides and angles in right-angled triangles, but when I asked them to explain why this worked, no one could explain using the concepts of similarity and ratios of sides. To my students, SOHCAHTOA were buttons on a calculator and a memorized procedure or algorithm to get an answer. These students had been exposed to traditional methods that focused on memorizing algorithms. Mathematics classrooms worldwide have tended to focus on rote learning procedures or algorithms, and often too little attention is paid to why or how. Why do we multiply by the reciprocal of the divisor when we divide fractions? Why can we not divide by zero? Why do two negatives make a positive? These are examples of important questions that need to be addressed to support conceptual understanding.

Waterbury School System in Connecticut embarked on the concept-based curriculum model in 2012. Darren Schwartz, the Instructional Leadership Director of Waterbury Public Schools, explains the reasoning behind adopting the concept-based approach:

> **T**he mission of the Waterbury School System is to establish itself as the leader in Connecticut for urban education reform in partnership with the State Department of Education and the entire Waterbury community. The school system will provide opportunities for all students to maximize their skills and talents in an atmosphere where teaching and learning flourish under the never-wavering belief that all students can be

exemplary students, while becoming respectful, responsible, productive citizens vital to our community.

There has been a long-standing tradition in math instruction to teach and model using algorithms first. In our district we focus firstly on the conceptual understanding of math and provide the opportunity for students to discover algorithms through an inquiry based learning process.

Darren Schwartz, Instructional Leadership Director
Waterbury Public Schools, Connecticut

Strategies are a number of skills that learners use in a methodical and systematic way to support learning.

FIGURE 2.3: THE STRUCTURE OF PROCESS FOR TRIGONOMETRY

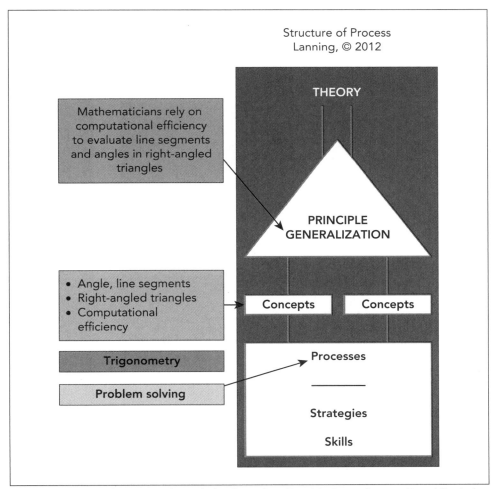

Adapted from original Structure of Knowledge figure from *Transitioning to Concept-Based Curriculum and Instruction*, Corwin Press Publishers, Thousand Oaks, CA.

Skills are small operations or actions that are embedded in strategies, and when appropriately applied, they "allow" the strategies to work. Skills underpin strategies. Examples of skills are being able to create a table of values, plot a graph, or to use trial and error to solve a problem (Erickson & Lanning, 2014).

Mathematical processes are complex, sophisticated performances, which are composed of strategies, algorithms, and skills (Erickson & Lanning, 2014). Mathematical processes are broad techniques that students draw upon when learning mathematics and support the understanding of the concepts in a unit of work. Problem solving is an example of a mathematical process.

Figure 2.3 (see page 31) looks at concepts related to trigonometry for a Structure of Process and how this relates to the problem solving process. **Problem solving** is a fundamental building block of mathematics and is an example of a sophisticated mathematical process.

Curricula around the world bring together core mathematical complex processes as a necessary component when learning math. These processes support conceptual understanding of process generalizations such as the example in Figure 2.3:

> **"M**athematicians rely on computational efficiency to evaluate line segments and angles in right-angled triangles."

Figure 2.4 compares the mathematical processes for different curricula from various countries around the world and some states in the United States. Note that this table does not represent a one-to-one correlation of each mathematical process however there is overlap and common themes. What is important to note is that each curriculum recognizes that mathematical processes are an important component when learning math and promote higher order thinking, developing reasoning, and critical thinking.

What do you notice about the common threads of mathematical processes between the different systems shown in Figure 2.4? The Common Core State Standards from the United States includes attention to precision and modeling, while Hong Kong and the International Baccalaureate include inquiry approaches and inquiry skills. The Australian Statements of Learning: Working Mathematically encompasses many processes, such as problem posing and solving, mathematical inquiry, representation, and communication.

Common to all systems is the importance of the mathematical processes of problem solving, reasoning, communicating mathematics in various ways, being able to recognize connections, and utilizing different tools to create representations. Although they are not described identically in every system, you can see how the processes align well.

1. Problem Solving
2. Reasoning and Proof
3. Communicating
4. Making Connections
5. Creating Representations

FIGURE 2.4: COMPARISON OF MATHEMATICS PROCESSES IN DIFFERENT PARTS OF THE WORLD

U.S. Common Core State Standards Mathematical Practices	Texas Essential Knowledge and Skills (TEKs) Mathematical Practice Standards	International Baccalaureate Assessment Objectives	Hong Kong Higher-Order Thinking Skills	Virginia Mathematics Standards of Learning	Australia Statements of Learning: Working Mathematically
SMP 1 Make sense of problems and persevere in solving them	Use a problem-solving model that incorporates analyzing given information, formulating a plan or strategy, determining a solution, justifying the solution, and evaluating the problem-solving process and the reasonableness of the solution	Problem solving	Problem solving	Mathematical problem solving	Working mathematically involves mathematical inquiry and its practical and theoretical application. This includes problem posing and solving, representation and modelling, investigating, conjecturing, reasoning and proof, and estimating and checking the reasonableness of results or outcomes. Key aspects of working mathematically,
SMP 2 Reason abstractly and quantitatively	Communicate mathematical ideas, reasoning, and their implications using multiple representations, including symbols, diagrams, graphs, and language, as appropriate	Reasoning	Reasoning	Mathematical reasoning	
SMP 3 Construct viable arguments and critique the reasoning of others	Display, explain, and justify mathematical ideas and arguments using precise mathematical language in written or oral communication	Communication and interpretation	Communicating skills	Mathematical communication	

(Continued)

U.S. Common Core State Standards Mathematical Practices	Texas Essential Knowledge and Skills (TEKs) Mathematical Practice Standards	International Baccalaureate Assessment Objectives	Hong Kong Higher Order Thinking Skills	Virginia Mathematics Standards of Learning	Australia Statements of Learning: Working Mathematically
SMP 4 Model with mathematics	Apply mathematics to problems arising in everyday life, society, and the workplace				individually and with others, are formulation, solution, interpretation, and communication. The processes of working mathematically draw upon and make connections between the knowledge, skills and understandings acquired in Number, Algebra, function and pattern, measurement, chance and data, and space.
SMP 5 Use appropriate tools strategically	Select tools, including real objects, manipulatives, paper and pencil, and technology as appropriate, to solve problems	Technology		Mathematical representations	
SMP 6 Attend to precision	Display, explain, and justify mathematical ideas and arguments using precise mathematical language in written or oral communication	Knowledge and understanding	Communicating skills	Mathematical Communication	
SMP 7 Look for and make use of structure	Analyze mathematical relationships to connect and communicate mathematical ideas	Knowledge and understanding	Conceptualizing skills	Mathematical connections / Mathematical representations	
SMP 8 Look for and express regularity in repeated reasoning	Analyze mathematical relationships to connect and communicate mathematical ideas	Inquiry approaches	Inquiry skills	Mathematical reasoning / Mathematical connections	

There is a sixth vital core mathematical process that is included in the International Baccalaureate and Hong Kong processes: the use of inquiry approaches and skills, which for the sake of ease I will call **investigating**.

These six complex mathematical processes cannot be completely isolated when learning mathematics or tackling a problem. For example, different representations may help you to reason and find a proof or communicate ideas. Each of these processes (represented on the Structure of Process) can be broken down into smaller strategies and skills.

Figure 2.5 is a model of the processes, algorithms, strategies, and skills for mathematics.

FIGURE 2.5: THE SIX MATHEMATICAL PROCESSES, ALGORITHMS, STRATEGIES, AND SKILLS

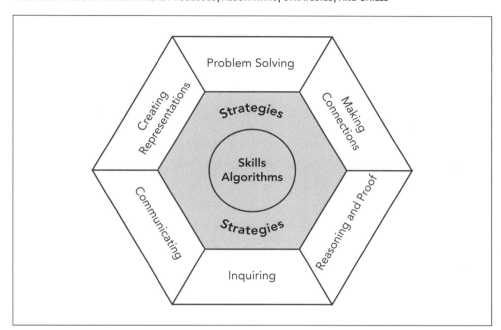

Transitioning to Concept-Based Curriculum and Instruction, Corwin Press Publishers, Thousand Oaks, CA.

Each mathematical process encourages higher order thinking and consists of different algorithms, strategies, and skills to be employed when learning math. Algorithms, strategies, and skills are usually specific to the unit of study, while these six mathematical processes are transferable across disciplines. For example, we can communicate, reason, and investigate in science or history.

Students need to know algorithms, such as how to substitute into a formula, multiply, or divide. An algorithm is specific to mathematics and is defined as a set of computational rules to produce a specified outcome. The order of operations rule PEMDAS (parentheses, exponent, multiplication, division, addition, and subtraction) is an example of an algorithm. (In the UK, the acronym BODMAS is used, which stands for brackets, order, divide, multiply, add, and subtract.)

Algorithms consist of skills, which allow learners to follow a set of rules leading to a specified outcome. Memorizing an algorithm does not necessarily imply conceptual

PEMDAS

Parentheses, Exponents, Multiplication and Division, Addition and Subtraction

To find the magnitude of a vector we use the Pythagorean theorem and follow this algorithm:

| 1. Substitute | 2. Evaluate the exponents | 3. Add them together | 4. Take the square root |

What is the magnitude of the vector **a** = 5**i**+ 12**j**?

$$|a| = \sqrt{5^2 + 12^2} \quad |a| = \sqrt{25 + 144} \quad |a| = \sqrt{169} \quad |a| = \sqrt{13}$$

understanding. Figure 2.6 illustrates an example of an algorithm for finding the magnitude of a vector (using the Pythagorean theorem) using the order of operations rule (PEMDAS or BODMAS).

Figure 2.7 lists some specific examples of skills and strategies for each of the six mathematical processes. Skills and algorithms form strategies, which support mathematical processes.

In mathematics, each of these six processes has specific meaning and applies to various situations. Most math learning involves employing a combination of two or more processes, so although they are interconnected, we can look at each process and its attributes in detail.

1. Problem Solving

The fundamental building block of mathematics is problem solving. It is a defined series of skills and strategies that form mathematical techniques that a learner employs when faced with an unfamiliar situation. Typical problem solving should be challenging, engaging, and pique learners' motivation and interest enough to encourage pursuit of the situation presented. Solving problems involves thinking logically and making decisions in a systematic way. George Pólya (1957, p. 253), a Hungarian mathematician, proposed the following principles when problem solving:

1. Understand the problem
2. Devise a plan
3. Carry out the plan
4. Review and extend

FIGURE 2.7: EXAMPLES OF SIX MATHEMATICAL PROCESSES BROKEN DOWN INTO SKILLS AND STRATEGIES

Problem Solving	Reasoning and Proof	Communicating	Making Connections	Creating Representations	Investigating
Act it out	Use data to make a conjecture	Listen as part of a group	Make connections between facts	Use words, drawings, tables, charts, graphs, diagrams, symbols, technology	Investigate unfamiliar situations
Interpret the question	Make generalizations	Discuss as part of a group			Explore different methodologies or pathways through questioning
Look for a pattern	Give justification for results	Write reflectively (e.g., what you have learned or journal)	Make connections between symbols and procedures	Include geometric, algebraic forms of representations	
Change your point of view	Ask "why"				Research and extract relevant ideas
Identify all possibilities	Recognize mistakes or flawed reasoning	Use correct mathematical terminology as a language	Make connections to the real world		Create different lines of inquiry
Guess and check	Use a variety of reasoning methods, such as	Question	Connect new problems to old		
Work backward	• Self-regulation	Criticize			
Write an open sentence	• Logical steps	Clarify			
Solve a simpler or similar problem	• Formulating and testing theories	Explain ideas			
Persevere	• Thinking critically				

Understanding the problem could mean rewording the question and extracting key ideas. At times, if one does not understand the problem, another strategy could be to look at a simpler or similar problem. Students need to develop a strong foundation in problem solving to understand concepts in math. Figure 2.7 gives examples of skills and strategies to use when problem solving. These include the ability to look for a pattern, identify all possibilities, guess and check, and interpret what the question is asking.

2. Reasoning and Proof

The ability to make generalizations and provide explanations and justification for arguments is key to the reasoning skill. Reasoning leads to a web of generalizations, which provide an interconnectedness of mathematical knowledge. Reasoning also leads to building mathematical memory to understand concepts. Learning through reasoning gives learners opportunities to make mistakes. Mistakes and misconceptions are a powerful tool in the learning cycle. The classroom environment needs to support, almost champion, the power of mistakes in learning. Critical thinking skills and asking "why" can drive the steps in the process of reasoning and proof.

Once students are guided through reasoning and proof to a generalization through inductive inquiry, mathematics affords us the beauty of proving these generalizations. Figure 2.8 is an alternative proof to the double angles identity of sin $2x$ and cos $2x$, which can be used after students discover through inductive inquiry the double angle results.

3. Communicating

Communicating refers to talking and writing about mathematics. It is important for learners to not only be given opportunities to explain and describe their ideas and stages in learning but also to be given opportunities to listen and discuss descriptions and explanations. Presenting different methodologies equips learners with alternative perspectives and enables learners to build a repertoire of strategies for problem solving. Discussion, sharing ideas, and collaboration are important aspects when learning any discipline. Ideally this is in the form of paired or small group work, rather than overreliance on teacher-led whole class discussion.

Number Talks

Number talks are based on the 1990s research of Kathy Richardson and Ruth Parker (Number Talks Tool Kit, 2015). This type of communication is especially important in laying the foundation of mental methods that help students understand higher, more sophisticated mathematics studied later. **Number talks** are a communication tool to help students develop computational fluency by comparing number relationships and looking at the different ways to add, subtract, multiply, and divide.

FIGURE 2.8: AN EXAMPLE OF REASONING AND PROOF

The Double Angle Results

ABC is an isosceles triangle with
AC = AB = 1

Draw a perpendicular line from apex A
to BC and call the intersection D. This
creates two right-angled triangles, ABD
and ACD.

Label angle CAD and BAD "x" on this
diagram.

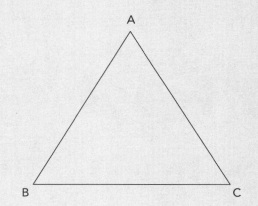

The expression for angle ACD in terms of x is _____

Similarly, angle ABD is _____

In triangle ACD, side AD is _____ and side CD is _____
because of SOHCAHTOA!

The sine rule is

Applying this to our original triangle, ABC, find an expression for sin 2x.

Using this triangle, find a simpler expression for sin (90 – x)

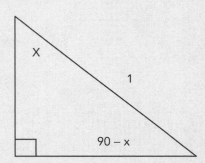

sin 2x =

The cosine rule is

Applying the cosine rule to triangle ABC, we have:

Which gives:

cos 2x =

Now, using $\sin^2 x + \cos^2 x = 1$, find two other forms of this double angle result.

cos 2x =
     =
     =

Explain what you have proved. What are the applications and significance of these identities?

For a completed version of Figure 2.8, please visit the companion website.

Just as we can communicate ideas using words, we can also communicate with each other using numbers and equations. Number talks are initiated by oral conversations. Written and graphical representations of mathematical solutions can help support a number talk. In order to ensure that students are learning to become (emerging) proficient mathematicians, we should teach them to communicate with numbers as well as words. Number talks are characterized by the following:

Class discussion

Classroom environment and community

The teacher's role

Mental mathematics

Purposeful computation problems

To establish effective foundations in numeracy, number talks should be regular and consist of 5 to 15 minutes of purposeful computation. I have started this practice with a Year 7 (Grade 6) class, and I use the template in Figure 2.9 to draw out different methods by asking students to communicate (both orally and in written form) their methodology.

My Year 7 (Grade 6) class has looked at different ways to mentally multiply two–digit numbers by one–digit numbers.

Figure 2.10 is an example of different methods from my very capable student, Bhuvan, who was asked to mentally work out 23×5. When students are able to see and create different visual, schematic representations of different methods, they lay a foundation for their future understanding in algebra of the associative and distributive properties to work out questions such as $3(x + 4)$ or even $(x + 3)(x - 2)$.

Research also shows that learners who have a variety of different methods to draw upon are more proficient in math. Bhuvan is an extremely able student who shows three different strategies for multiplying 23 by 5. This type of activity provides a foundation for algebraic techniques in the future such as using the distributive property, understanding inverse operations, and using the associative property.

Engaging in number talks supports students' communication of mathematics and allows them to create different representations of various strategies to employ in the future.

Social Constructivist Approach

Another form of communication is the social constructivist approach. The Russian psychologist Lev Vygotsky (1896–1934) laid the foundation for social constructivist theory, which states people construct their own understanding and knowledge of the world through experiences, reflecting on those experiences, and collaboration among learners. Learners cognitively develop in a context of socialization (Vygotsky, 1978).

FIGURE 2.9: NUMBER TALKS TEMPLATE

Number Talks

In pairs, you will be discussing and writing down some methods you used *in your head* to work out the question posed.

Question:

Mentally work the answer out and write down the answer here:

Write down your method here below. Can you represent your method in a diagram?

Explain to your partner how you worked out the answer to the question posed.

Write down your partner's method here:

As a pair, make up another question, but you now must use your partner's method to work out the answer in your head.

Think of three other ways that you could work out the answer to your question and explain the methods here.

Instruction and culture are also important for knowledge acquisition, and Vygotsky asserted three main principles, which I have applied to learning mathematics:

1. Children learn effectively from one another.

2. Mathematics should be incorporated into meaningful tasks with purpose.

3. Discussion is an important element in the internalization of mathematical ideas and understanding.

Social constructivism emphasizes assisted discovery through teacher-student and student-student interactions. This implies learners need to be given opportunities to communicate with each other. Discussion helps learners to form and construct mental structures for deep conceptual understanding.

In the number talk activity, students discuss their methods and see different schematic representations of the problem. This helps students to construct their own understanding of the concepts of associative and distributive properties.

Uri Treisman (1992) researched the disparity between Asians and other minority (Latino and black) races in terms of student achievement in calculus. Both groups had similar homogeneity in terms of socioeconomic levels and background knowledge. Why were the Asians performing better than other minority races? Treisman found Asians regularly met socially to discuss the work and compare answers; the Asians learned from each other. They discussed, debated, and edited each other's solutions. In contrast, the other minority groups worked independently and often in isolation. Treisman introduced the idea of intensive workshops to supplement the calculus course, giving students opportunities to collaborate, work in groups, and learn from each other through discussion. The minority groups who had enrolled in these workshops not only outperformed their own minority peers but also white and Asian students, regardless of the differences in background knowledge or SAT scores. These workshops provided learning in a social context, which promoted reasoning and conceptual growth. Due to the effectiveness of this approach, most university undergraduate and postgraduate programs now incorporate group projects as part of assessment to encourage learning through collaboration.

Writing Mathematics

Another strategy to encourage students to communicate their conceptual understanding is writing mathematics. David Sousa's studies on the brain reveal that writing mathematics, in the form of words and explanations, helps students to learn mathematical concepts more effectively, encouraging higher order thinking skills. Writing mathematics creates a permanent record of learners' thoughts and how they organized their ideas to develop new applications and to solve problems. David Sousa (2015) explains, "In addition to requiring focus, writing provides another modality for processing information and skills, thereby helping the students to find sense and meaning, and increasing the likelihood that the new learning will be remembered" (p. 62).

In the Double Angle reasoning and proof worksheet (Figure 2.8), students were asked the following questions to encourage the writing of mathematics:

Explain what you have proved.

What are the applications and significance of these identities?

These types of prompts encourage students to express their understanding in the form of prose. Other examples of writing prose could be writing in a reflection journal or putting together written summaries after a unit to reinforce the concepts covered.

Different modes of communication allow students to develop communication skills. These include reading and writing correct mathematical terminology and, of course, nonverbal and verbal forms of communication. If students do not learn to discourse using the language of mathematics—which is conceptual—then they will fail to demonstrate mathematical literacy.

4. Making Connections

Mathematics is a cumulative body of knowledge that has well-integrated connectedness. The process of **making connections** refers to the learner's ability

- To see connections between facts and how they relate to one another;
- To make connections between symbols and procedures;
- To make connections between what they are learning and the real world;
- To connect new problems to old; to see different concepts and how they relate to one another.

A graphic organizer can help students to form the connections within a unit of work. Figure 2.11 is a graphic organizer showing how rational numbers, real numbers, natural numbers, and integers relate to each other in a Venn diagram. An extension of this organizer would include how complex numbers and imaginary numbers relate to the other sets in the diagram. By showing students a graphic organizer such as Figure 2.11, or, better yet, asking them to fill out a blank one themselves, you are using another means of communication (visual) to demonstrate the connections of different sets of numbers. If students are able to accurately fill out the Venn diagram themselves, they are approaching a better understanding of the concepts being taught.

5. Creating Representations

Representations (such as Figure 2.11) are ways to express different mathematical ideas; they include graphs, tables, geometric figures, and so on. In the previous discussion on number talks (Figure 2.10), Bhuvan was asked to communicate verbally and nonverbally and also asked to create different representations. Number talks are a combination of communicating and creating representations processes. Different

FIGURE 2.11: AN EXAMPLE OF MAKING CONNECTIONS AND CREATING REPRESENTATIONS: GRAPHIC ORGANIZER FOR THE NUMBER SYSTEM

The Number System

Place the correct type of number in the Venn diagram

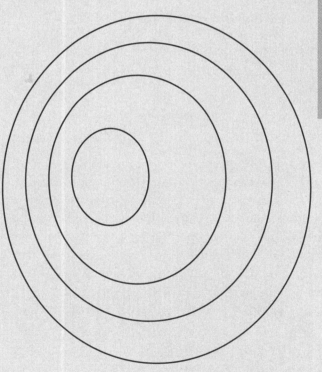

$\mathbb{N}$ represents the set of all natural numbers (positive integers)

$\mathbb{Z}$ represents the set of all integers

$\mathbb{Q}$ represents the set of all rational numbers

$\mathbb{R}$ represents the set of all real numbers

For a completed version of Figure 2.11, please visit the companion website.

representations can help learners to understand in different ways and see connections. The process of creating mathematical representations aligns with one of the Effective Mathematics Teacher Practices of the National Council of Teachers of Mathematics (2014) which states, "Use and connect mathematical representations. Effective teaching of mathematics engages students in making connections among mathematical representations to deepen understanding of mathematics concepts and procedures and as tools for problem solving."

The purpose of representation is threefold:

1. To help learners organize, record, and communicate mathematical ideas;

2. To select, apply, and translate different forms of representation to solve problems; and

3. To use representations to model and interpret physical, social, and mathematical phenomena.

The process of constructing various representations of the same mathematical object leads to deep understanding and enhances the problem solving process. As an example, students understanding that there are five equivalent representations of the same linear equation with two variables and being able to, when starting from any of them, create the remaining, helps them choose the most appropriate one when unfamiliar situations are given (Diagram A).

DIAGRAM A

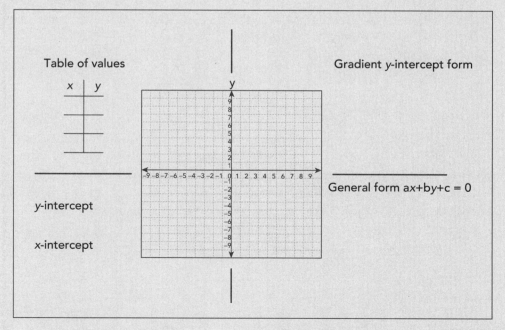

(Continued)

(Continued)

More than that, a new, related concept (e.g., growing patterns) will fit into this diagram and students will naturally create the conceptual bridge between the two (Diagram B).

DIAGRAM B

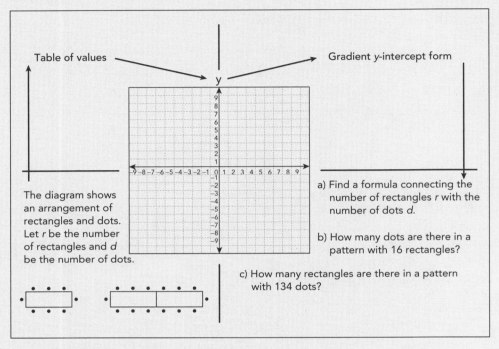

Dr. Daniela Vasile, Head of Mathematics
South Island School, Hong Kong

Mathematics should be represented visually at every opportunity to help students to understand the concepts in learning, regardless of whether a learner is a visual learner. The use of technology can create different representations, such as using regression models to model real-life situations. In addition to helping your students understand the material, instilling in them the habit of creating visual representations themselves is more likely to lead to their own academic achievement in mathematics.

Montague and Van Garderen (2003) concluded from their study that gifted math students utilized significantly more visual spatial representations when problem solving. According to them, "Successful mathematical problem solving was positively correlated with use of schematic representations; conversely, it was negatively correlated with use of pictorial representations." (p. 246). The difference between pictorial representations and schematic representations is that **pictorial repre-sentations** include drawings that contain irrelevant aspects of the problem, such as drawing a picture of the scenery of a problem and coloring this in. In contrast, **schematic representations** are purposeful to solving the problem and are codes or symbols to show relationships and connections.

One useful way for students to practice schematic representations is to ask students to employ the grid method to a problem. This can be done at the elementary, middle, or high school level.

Figure 2.12 below shows three examples of how the grid method can be applied to problems of various degrees of complexity. Using the grid method can simplify a problem considerably by creating a visual representation. It allows a student to see and use a different representation, allowing the brain to approach the problem from a new angle.

The idea behind the grid method is to utilize the distributive property when multiplying two-digit numbers and binomials. The grid method facilitates understanding by creating a representation of the problem. The process of completing the grid method involves breaking down a complex problem into a more straightforward multiplication sum. For the elementary example, two-digit numbers are split into tens and units (e.g., $23 = 20 + 3$) and represented in a grid. This progresses to separating a binomial at middle school and then solving a more sophisticated problem involving binomial expansion at high school level.

6. Investigating

Investigating means being able to explore unfamiliar mathematical situations. When investigating, you ask questions of yourself in order to further your mathematical understanding. This also includes the ability to be resourceful by researching information and extracting relevant ideas from your research. It is important that curriculum and instruction promote creativity when approaching difficult and unfamiliar math problems by encouraging students to investigate, probe, ask questions, try different approaches, and conduct their own research. By encouraging investigating in mathematics, teachers allow students to experience the beauty of mathematics, which is grounded in creative expression. Andrew Blair's website (http://www.inquirymaths.com) includes numerous tasks that encourage the process of investigating.

The Iterative Stages When Learning Mathematics

When students employ processes, strategies, and skills, there are iterative stages due to the cumulative nature of mathematics. Learning math involves the input of processes, strategies, and skills to support the understanding of concepts. Each iteration

FIGURE 2.12: THE DIFFERENT USES OF THE GRID METHOD

Example 1: Grid Method at Primary Level

The Grid Method

Work out: 23 × 41 using the grid method:

	20	3
40		
1		

Adding all four rectangles gives:

Example 2: Grid Method at Middle School Level

Expand $(x + 3)(x - 2)$ using the grid method.

	x	3
x		
−2		

Adding all four rectangles gives:

Example 3: Grid Method at High School Level

Grid Method for Binomial Expansions

Question:

Find n in the following

$$(1+\frac{2}{3}x)^n(3+nx)^2 = 9+84x+....$$

Using the grid method simplifies the problem considerably by giving you a visual approach; a different representation for the product of two binomials.

Guidance:

We are going to expand each bracket separately first:

1. $(1+\frac{2}{3}x)^n = 1+ {}^nC_1\frac{2}{3}x+...$

Where ${}^nC_1 = $ _____

Do I need any more terms? Why or why not?

2. $(3+nx)^2 = $ _____

3. Which boxes do I need to help me find n?

	1	${}^nC_1\frac{2}{3}x$	
9			
6nx			

Now equate the coefficients of x to find n.

  For a completed version of Figure 2.12, please visit the companion website.

forms conceptual understandings through continuous stages. Processes, strategies, and skills in math lead to understanding of concepts, which then leads to more strategies and skills.

Macro, Meso, and Micro Concepts in Mathematics

Recall from Chapter 1 that features of concepts are mental constructs—organizing ideas framed by common attributes.

In summary, concepts meet the following criteria:

- Timeless
- Universal
- Transferable
- Abstract to varying degrees
- One or two words, or a short phrase
- Have different examples that share common attributes (Erickson, 2007)

Mathematics is inherently a language of conceptual relationships. As explained by Erickson and Lanning (2014), "Mathematics is structured by its concepts and the conceptual relationships; the examples are the supporting facts. . . . For example concepts such as subtraction or algebraic equations are supported by specific algorithms. So we consider mathematics primarily a concept driven discipline" (p. 40).

> Macro concepts give us breadth of understanding in math, and micro concepts give us the depth of understanding.

As mathematics is a conceptual language, it is logical to categorize its concepts as macro, meso, or micro. **Macro concepts** give *breadth of understanding* and possess the greatest transferability across *most* disciplines. In mathematics, however, we consider macro concepts our broader constructs that encompass a huge body of knowledge (e.g., geometry, algebra, and calculus). These macro concepts can be further broken down into **meso concepts** (e.g., trigonometry) and **micro concepts** (e.g., ratio, angles, and similarity). Micro concepts give us the supportive *depth of understanding*.

For example, understanding the micro concepts of magnitude, direction, and angle help us to focus our understanding of the meso concept of vectors, which falls under the umbrella of geometry. Figure 2.13 shows some examples of macro, meso, and micro concepts in mathematics.

FIGURE 2.13: EXAMPLES OF MACRO, MESO, AND MICRO CONCEPTS IN MATHEMATICS

Macro Concepts	Meso Concepts	Micro Concepts
calculus	differentiation, integration	slope, rate of change, velocity, limits, derivative, summation
algebra, algebraic expressions	functions	quadratics, arithmetic sequences, geometric sequences, complex numbers, variables and symbols, linear function, xy-plane, domain, range, graphs, inverse functions, composite functions
statistics and probability	bivariate analysis, univariate analysis	relative frequency, regression, central tendency, dispersion
number	measurement	multiplication, estimation, patterns, prime, factor pair, triangular numbers, square numbers, magnitude
geometry	trigonometry, vectors	ratio, radian measure, direction, magnitude, unit vectors, base vectors, scalar product, vector product, angles, polygons

Because mathematics is a conceptual language, all topics in the Structure of Knowledge are concepts. Figure 2.14 outlines some of the concepts in three of the IB diploma math courses. This table gives examples of micro concepts involved when learning math in high school.

The Marriage of the Structure of Knowledge and the Structure of Process

The interplay between the Structure of Knowledge and the Structure of Process is relevant in mathematics and refers to the knowledge and the strategies and skills (which form processes) needed for mathematical understanding. Key skills and processes are what students will be able to *do* (e.g., plot points on a Cartesian plane). *Knowledge* refers to the content, such as the facts and formulae. Erickson and Lanning (2014) describe the relationship between process and knowledge as complimentary and symbiotic (p. 26).

Figure 2.15 illustrates how the processes, made up of concrete algorithms, strategies, and skills along with facts, build on the conceptual level.

FIGURE 2.14: EXAMPLES OF MICRO CONCEPTS IN THE INTERNATIONAL BACCALAUREATE DIPLOMA MATH COURSES

IB Mathematics (Standard level)	IB Mathematics (Standard level)	IB Mathematics (Higher level)

patterns and sequences
mathematical models
standard units
non-standard units
estimation
converting units
units of measure
linear functions
variables and symbols
algebraic expressions
order of operations
distributive property
coordinate plane
graphs
linear parent functions
simultaneous equations
	independent/
	   dependent
	variables
	functional relationships
	domain/range
	patterns and sequences
	discriminant
	quadratic functions
	logarithmic &
	   exponential
	   functions
	trigonometric functions
	trigonometric identities
	gradient/slope
	rate of change
	summation
	correlation
	scatter plots
		patterns and sequences
		discriminant
		independent/dependent
		variables
		functional relationships
		domain/range
		quadratic functions
		logarithmic &
		   exponential functions
		roots, zeros of
		   polynomials
		induction
		equations of planes
		combinations and
		   permutations

When learning mathematics, it is the marriage of the content in the Structure of Knowledge and the processes in the Structure of Process that support deep conceptual understanding. Imagine learning about

- $y = mx + c$, $y = ax^2 + bx + c$, but not knowing how to plot a set of coordinates or put together a table of values;

- the formulae for arithmetic sequences, but not being able to subtract, add, solve an equation, or problem solve;

- right-angled trigonometry, but not knowing how to substitute, multiply, or work backwards.

Creating a synergy between the facts, skills, and conceptual understanding is at the heart of concept-based curriculum. In the words of Lynn Erickson (2007), "To stimulate more sophisticated, complex thinking we need to create a *synergy* between the simpler and more complex processing centers in the brain. This interactive synergy requires the mind to process information on two cognitive levels—the factual and the conceptual. The conceptual mind used facts as a tool to discern patterns, connections and deeper, transferable understandings" (p. 10).

This synergistic relationship can be cultivated through the use of a conceptual lens, generalizations, and guiding questions, which will be discussed in Part II of

FIGURE 2.15: HOW PROCESSES, SKILLS, AND FACTS CONTRIBUTE TO AN UNDERSTANDING OF CONCEPTS

Macro Concepts	Meso Concepts	Micro Concepts	Examples of Facts	Processes	Examples of Algorithms, Strategies, and Skills
calculus	differentiation integration	slope, rate of change, velocity, limits, derivative, summation	$\dfrac{dy}{dx} = nx^{n-1}$	problem solving reasoning communicating creating representations making connections investigating	substitute into a formula, rearranging (following set of rules), finding specified output
algebra	functions matrices	quadratics, variables, formula linear function	$y = ax^2 + bx + c$ $y = mx + c$ $x = \dfrac{-b \pm \sqrt{b^2 - 4ac}}{2a}$		extracting variables from word problems, algebraic coding plotting points balancing and solving equations
statistics	bivariate analysis univariate analysis	regression	$\bar{y} = a + bx$		analytical skills, plotting a scatter graph
number	measurement	multiplication, estimation	1 mile = 1.6 km		numeracy skills: add, subtract, multiply, divide measure objects with a ruler
geometry	trigonometry vectors	angles	$\sin x = \dfrac{opposite}{hypotenuse}$		graphical and algebraic skills for solving trig equations use a protractor to measure

this book: Chapters 3 and 4. A **conceptual lens** sets up a synergy between the factual and the conceptual processing centers of the brain. The conceptual lens is a broad, integrating concept that focuses a unit of work to allow students to process the factual information. Examples of conceptual lenses include change (this could be used when we study differentiation in calculus), accumulation, structure, form, or order.

Figure 2.16 breaks down the macro concept of geometry into the meso concept of trigonometry from the Structure of Knowledge and the Structure of Process perspective. This figure illustrates the dual part the Structure of Knowledge and the Structure of Process play in the development of mathematical understanding and, ultimately, intellect.

Two-Dimensional vs. Three-Dimensional Curriculum Models

The lower layers of the Structures of Knowledge and Process represent a two-dimensional model that traditional curriculum and instruction has focused on. This results in a **transmission model of instruction** by which teachers "transmit" knowledge for students to regurgitate in tests. In the two-dimensional model, there has also been an overemphasis on skills and algorithms, with little conceptual understanding. The lower layers of the Structure of Knowledge and the Structure of Process include both facts (content) and skills and exist in a symbiotic relationship in math learning; one cannot fully thrive without the other. Processes are complex skills and strategies and encompass what we want our students to be able to do. At the same time, learners have to know facts and content to help build their conceptual understandings. What we want students to know, understand, and do provides critical content for unit planning, which is discussed in Chapter 4.

> Our KUDs (what we want our students to know, understand, and do) provide the critical content when unit planning

Let us look at another example in mathematics. The function $f(x) = x^2 - 3$ can be understood or analyzed using both the Structure of Process and the Structure of Knowledge. Looking at the function using both approaches offers a fuller understanding of the meso concept of function.

Structure of Process. If we look at $f(x) = x^2 - 3$ using the Structure of Process, we ask students to substitute values of x to calculate the value of the function. Students would also be expected to represent the values of x in different forms, such as graphs or using a table of values, as shown in Figure 2.17.

Structure of Knowledge. If we look at $f(x) = x^2 - 3$ using the Structure of Knowledge, we ask students to be able to explain (both in prose and verbally) the definition of a function in terms of both one-to-one and many-to-one mappings. They would

FIGURE 2.16: STRUCTURES EXAMPLE SIDE-BY-SIDE FOR THE MESO CONCEPT TRIGONOMETRY

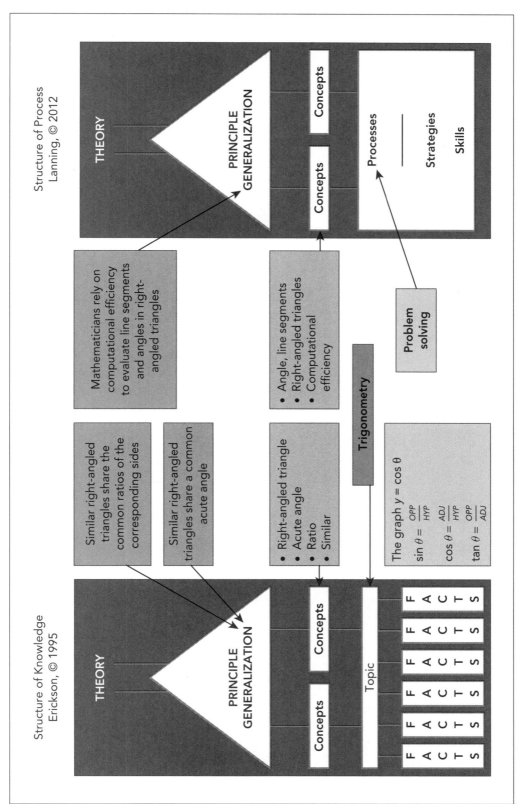

Adapted from original Structure of Knowledge and Structure of Process figures from *Transitioning to Concept-Based Curriculum and Instruction*, Corwin Press Publishers, Thousand Oaks, CA.

also be expected to understand that $f(x) = x^2 - 3$ is a quadratic function. The Structure of Knowledge approach would expect that students understand and are able to define, in their own words, terms such as *function, mappings, quadratic, parabola,* and *transformation*.

FIGURE 2.17: TABLE OF VALUES

x	−1	0	1	2	3
$f(x)$	−2	−3	−2	1	6

The micro concepts that we wish our students to understand regarding the function $f(x) = x^2 - 3$ are function, mappings, quadratic, parabola, substitution, and transformation. The Structure of Knowledge and the Structure of Process marry to support the conceptual understanding of functions.

Concepts, principles, and generalizations help our students to understand the relevance of facts and skills in context and move curriculum and instruction toward a three-dimensional model. The three-dimensional model of concept-based curriculum includes the conceptual level, which suggests a more sophisticated design for curriculum and instruction. The three-dimensional model allows students to internalize the material on a deeper level that can be likened to Daniel Pink's (2005) description of *High Concept and High Touch.* "High Concept involves the capacity to detect patterns and opportunities, to create artistic and emotional beauty, to craft a satisfying narrative, and to combine seemingly unrelated ideas into something new" (p. 2).

A two-dimensional model of curriculum and instruction served the Industrial Age well; however, does this model prepare our students for the Conceptual Age?

Although mathematics *can* be taught from a purely content-driven or process-driven perspective, using one or the other often fails to ensure that students attain conceptual understandings of the material. Ideally, it is the marriage of the two—content and processes—that promotes deeper conceptual understanding.

Chapter Summary

This chapter discussed the levels in the Structure of Knowledge and the Structure of Process for mathematics, with clear explanations of the terms *facts, topics, skills, algorithms,* and *concepts.* The Structure of Knowledge for the mathematics realm has specific meaning. The facts in the Structure of Knowledge are the memorized definitions, vocabulary, and formulae. Topics in the Structure of Knowledge are actually broader concepts—meso concepts—that cover a wide body of knowledge and give us breadth of understanding. The meso concepts break down into more specific micro concepts at the next level in the Structure of Knowledge, which gives us depth in our understanding. Generalizations and principles are statements of conceptual relationship. Principles include theorems in mathematics. The Structure of Process relates the concepts drawn from mathematical processes, strategies, and skills to help students understand process-driven conceptual understandings. Mathematical processes around the world align with the following categories:

1. Problem solving
2. Reasoning and proof
3. Communicating
4. Making connections
5. Creating representations
6. Investigating

These processes encourage higher order thinking and cannot be completely isolated when learning math. Most math learning involves employing a combination of two or more processes. Each process is discussed in detail and examples of strategies and skills that form the mathematical processes is provided.

The Structure of Knowledge and the Structure of Process are complementary, three-dimensional, concept-based models that have a symbiotic relationship. Curriculum and instruction has traditionally focused on the lower levels of the Structure of Knowledge and the Structure of Process, resulting in a two-dimensional model. The three-dimensional concept-based model incorporates the conceptual level and includes the higher levels in the Structure of Knowledge and the Structure of Process. The synergistic relationship between the factual and conceptual levels of knowing and understanding is key to a concept-based curriculum and to developing intellect. In order for deeper mathematical thinking and understanding, we need to focus on cultivating and nurturing conceptual understandings through an understanding of the Structure of Knowledge and the Structure of Process. How to craft statements of conceptual understanding, generalizations, and how to plan concept-based units of work will be discussed in the next part of the book.

Discussion Questions

1. What are facts in mathematics?

2. What is the difference between formulae and generalizations in mathematics?

3. What is the distinction between processes, algorithms, and skills in mathematics?

4. What are the key categories of processes in mathematics? Provide examples for each process.

5. How do the Structure of Knowledge and the Structure of Process represent different facets of learning mathematics? Explain the symbiotic relationship between the two structures for learning mathematics.

6. How is mathematics a language of conceptual relationships made of macro, meso, and micro concepts?

7. Where on the Structure of Knowledge and the Structure of Process should curriculum and instruction focus? Why?

Part II

How to Craft Generalizations and Plan Units of Work to Ensure Deep Conceptual Understanding

CHAPTER 3

What Are Generalizations in Mathematics?

What is π? Often students cannot explain what π is, stating it is 3.14 or 22/7. In fact, π is the ratio of the circumference to the diameter of *every and any* circle. What are the trigonometric ratios? A common explanation is that trigonometric ratios are buttons on a calculator, or they are sine, cosine, or tangent of an angle as formulae to be memorized. In fact, these three trigonometric ratios connect ratios of sides of similar right-angled triangles, which help to solve real-life problems in surveying, architecture, and astronomy.

Generalizations give us explanations of why and what we want our students to comprehend in terms of the relationship between two or more concepts. Also known in education circles as *enduring understandings, essential understandings,* or *big ideas* of the unit, they summarize what we would like our students to take away after their unit of study. Liz Bills and colleagues (2007) explain, "Provoking generalisation is more about releasing learners' natural powers than it is about trying to force feed. Because promoting mathematical generalisation lies at the core of all mathematics teaching, at all ages, and because it concerns the development of higher psychological processes that are most likely to be accessible to learners if they are in the presence of someone more expert displaying disposition to and techniques for generalising, it is important for teachers to be seen to generalise, to value learner's attempts at generalisation, and to get out of the learner's way so that they can generalise for themselves" (p. 56).

Generalizations provide clear goals and align with the National Council of Teachers of Mathematics (NCTM, 2014) Mathematics Teaching Practices:

> Establish mathematics goals to focus learning. Effective teaching of mathematics establishes clear goals for the mathematics that students are

learning, situates goals within learning progressions, and uses the goals to guide instructional decisions.

What Is the Difference Between a Generalization and a Principle in Mathematics?

Lynn Erickson (2014) distinguishes between generalizations and principles in mathematics by explaining that "Generalizations are defined as two or more concepts stated as a sentence of relationship. They are understandings that transfer through time, across cultures, and across situations. Generalizations contain no proper nouns, past tense verbs, or pronouns that would associate the idea with a particular person or group .... Generalizations are truths supported by factual examples, but they may include a qualifier (*often, can, may*) when the idea is important but does not hold in *all* instances" (p. 34).

Generalizations are the overriding ideas we would like our students to understand and transfer that may contain a qualifier such as *may* or *can*. Because mathematics is a conceptual language, generalizations are statements of two or more concepts that transfer through time, cultures, and across situations in the context of mathematics. For example, the angle sum in triangles *may* add to 180 degrees. In Euclidean geometry (the study of "flat space"), triangles always add up to 180 degrees. However, if we look at non-Euclidean geometry, this generalization does not hold true, so the qualifier *may* is used. Mathematical generalizations can also be expressed with an appropriate domain, so a qualifier is not needed.

Generalizations provide goals when unit planning and help learners to construct and make connections with what they are learning. Through inductive, inquiry-led teaching, learners are guided to discover generalizations and answer the question, "Why am I learning this and what is the point beyond this example?"

Wiggins and McTighe (2006a) talk similarly about statements of understanding, which they refer to as "enduring understandings": "In UbD (Understanding By Design), designers are encouraged to write [enduring understandings] as full sentence statements, describing what, specifically, students should understand about the topic. The stem "Student will understand that . . ." provides a practical tool for identifying these understandings" (p. 342).

The term *enduring understandings* is the same as Generalizations and Principles in the Structures of Knowledge and Process. There are many terms that different schools use to identify these statements of conceptual relationship: generalizations, essential understandings, enduring understandings, or statements of inquiry. Wiggins and McTighe (2006b) define two types of enduring understandings: *overarching* and *topical*.

Overarching enduring understandings (**overarching generalizations**) are understandings beyond the specifics of a unit. They are the larger, transferable insights we want students to acquire. They often reflect year-long program understandings.

Topical understandings (**topical generalizations**) are subject or topic specific. They focus on a particular insight we want students to acquire within a unit of study. Figure 3.1 is an example of overarching and topical enduring understandings, adapted from Wiggins and McTighe (2006b, p. 114).

This chapter will discuss topical enduring understandings, which will be referred to as *topical generalizations,* and how to craft topical generalizations for mathematics.

Principles in Structures of Knowledge and Process are laws or foundational truths that hold all the attributes of generalizations and commonly describe real-life situations. Theorems are the "principles" in mathematics. Principles never contain qualifiers such as *may, can,* or *often.* Mathematical theorems, such as the Pythagorean theorem, Fermat's theorem, and Pick's theorem, are all principles when written out as statements of relationship in either the Structure of Knowledge or the Structure of Process. Some theorems are derived from observations around the world and are considered the laws of mathematics.

FIGURE 3.1: THE TWO TYPES OF ENDURING UNDERSTANDINGS: OVERARCHING AND TOPICAL

Overarching Enduring Understanding	Topical Enduring Understanding
Mathematics reveals patterns that might have remained unseen.	Statistical analysis and graphic displays often reveal patterns in seemingly random data or populations, enabling predictions.

Examples of Mathematics Generalizations

Let's look at some specific examples of mathematics generalizations and principles and how they are supported by the concepts of the unit, processes, and facts.

Generalization Example 1

Utilizing algebraic tools such algebraic multiplication, subtraction, and division allow highly complex problems to be solved and displayed.

The concepts of *algebraic tools, multiplication, subtraction,* and *division* support the preceding generalization. This is a process generalization that supports the process of problem solving and can be represented in the Structure of Process as shown in Figure 3.2.

FIGURE 3.2: THE STRUCTURE OF PROCESS FOR EQUATIONS

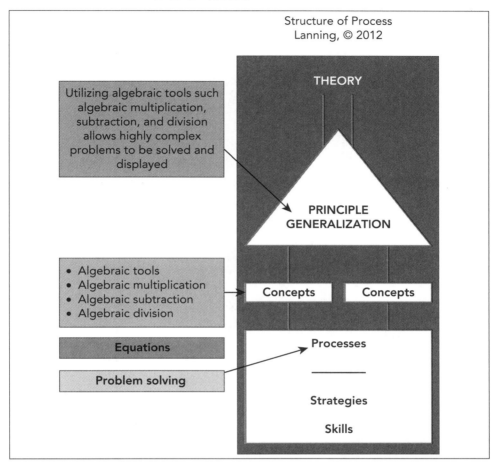

Adapted from original Structure of Process figure from *Transitioning to Concept-Based Curriculum and Instruction,* Corwin Press Publishers, Thousand Oaks, CA.

Generalization Example 2

The following generalizations refer to the meso concept *vectors.* The scalar product is used to work out magnitudes and angles between two vectors.

The scalar product between vectors represents the angle between the vectors and magnitude of the vectors.

Scalar product determines perpendicular and parallel vectors by projecting component vectors.

These generalizations are supported by the fact $a \cdot b = |a|\,|b|\cos\theta$ and are supported by the concepts of *magnitude, direction, angle, perpendicular, parallel,* and *component vector,* which can be represented diagrammatically in the Structure of Knowledge, as shown in Figure 3.3.

FIGURE 3.3: THE STRUCTURE OF KNOWLEDGE FOR VECTORS

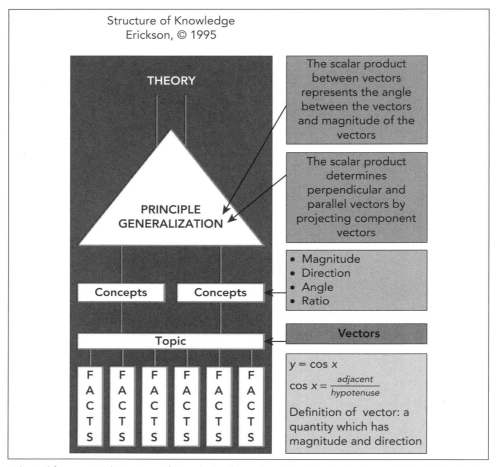

Adapted from original Structure of Knowledge figure from *Transitioning to Concept-Based Curriculum and Instruction,* Corwin Press Publishers, Thousand Oaks, CA.

Generalization Example 3

Many students find the concept of logarithms challenging. Students often memorize logarithmic laws with no understanding of their purpose. Understanding that logarithms are inverses of exponential functions and that exponential functions represent continuous compounded growth shows deep intellectual depth.

Logarithm laws give a means of changing multiplicative processes into additive processes, and this can provide the means to find inverses of exponential functions, which represent continuous compounded growth.

The two mathematical processes, *problem solving* and *making connections,* support this process generalization, as shown in Figure 3.4.

FIGURE 3.4: THE STRUCTURE OF PROCESS FOR LOGARITHMS

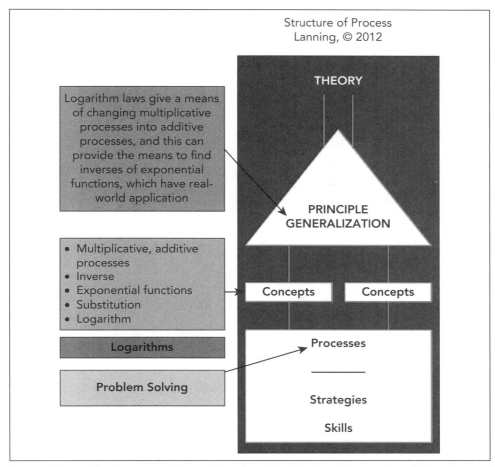

Adapted from original Structure of Process figure from *Transitioning to Concept-Based Curriculum and Instruction,* Corwin Press Publishers, Thousand Oaks, CA.

Generalization Example 4

The following generalizations summarize the concepts and significance of *finding the roots* (using the quadratic formula) and *discriminant* in quadratic equations.

The expression underneath the square root in the quadratic formula, the discriminant, determines the nature of the roots, which highlight geometrical features.

The quadratic formula determines the zeros of functions, the roots of equations, and the x-intercepts graphically.

These generalizations are supported by the following fact (mathematical formula):

For $ax^2 + bx + c = 0$

$$x = \frac{-b \pm \sqrt{b^2 - 4ac}}{2a}$$

These generalizations support the concepts of *equation, zeros, roots,* and *quadratic function* and can be represented in the Structure of Knowledge as shown in Figure 3.5.

FIGURE 3.5: THE STRUCTURE OF KNOWLEDGE FOR QUADRATICS

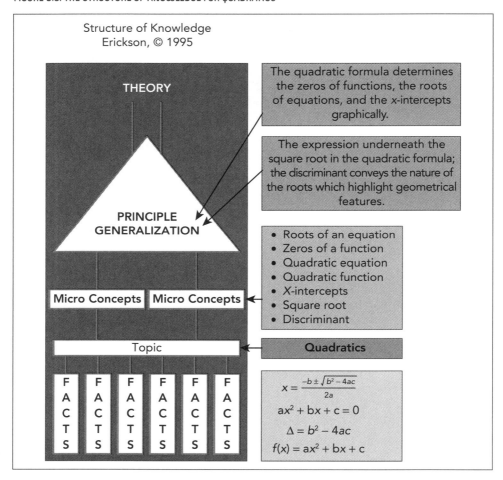

Adapted from original Structure of Knowledge figure from *Transitioning to Concept-Based Curriculum and Instruction*, Corwin Press Publishers, Thousand Oaks, CA.

Generalization Example 5

Figure 3.6 depicts the Structure of Knowledge and Structure of Process side by side for the meso concept *quadratics.* The micro concepts on the Structure of Knowledge are *equation, zeros, roots,* and *quadratic function.* These micro concepts support these two generalizations:

The quadratic formula determines the zeros of equations, roots of functions, and the *x*-intercepts graphically.

The expression underneath the square root in the quadratic formula, the discriminant, conveys the nature of the roots, which highlight geometrical features.

For the Structure of Process, the micro concepts are drawn from the process of creating representations in addition to following strategies and skills, such as using an *xy*-plane, using a table or graphs, and utilizing substitution. These micro concepts support the following generalization:

Creating visual depictions of a problem using different modes of representation (graphs, tables, etc.) helps explain the problem and reveal a solution.

The interplay between both these structures support learning and conceptual understanding in this unit. In other words, these structures support the understanding of the generalizations of a unit.

How Do We Craft Quality Mathematics Generalizations?

Utilizing quality generalizations when planning a unit of work guides the work of a concept-based teacher. Generalizations and principles in unit plans provide a clear focus for curriculum and instruction. Giving plenty of opportunities for students to generalize is the essence to learning mathematics. All concepts in mathematics can be expressed in a sentence of relationship in a variety of ways, according to factors such as student level and the teacher's aim. Because of this, the writing of generalizations should be a collaborative process within a faculty and part of the planning for a concept-based unit. A unit of work may typically have six to eight generalizations, with five to six instructional units over a year.

When starting to write generalizations, it is common to write statements that are too general. The verbs *is, are, have, impacts, affects* and *influences* are overused and create weak statements.

> Avoiding certain "no no" verbs will create stronger statements of conceptual understanding.

Lynn Erickson (2007) provides a three-step guide to help scaffold the process of writing quality generalizations and advises avoiding the use of the verbs *is, are, have, impacts, affects,* and *influences,* referring to these as "no no verbs."

FIGURE 3.6: SIDE BY SIDE: THE STRUCTURE OF KNOWLEDGE AND THE STRUCTURE OF PROCESS FOR QUADRATICS

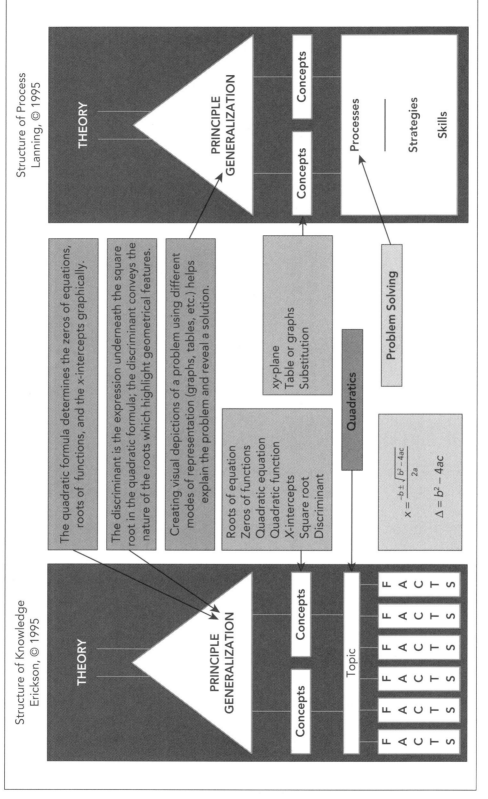

Adapted from original Structure of Knowledge and Structure of Process figures from *Transitioning to Concept-Based Curriculum and Instruction*, Corwin Press Publishers, Thousand Oaks, CA.

On the companion website you will find a table of "no no" verbs (Figure M2.1), two examples of how to scaffold generalizations (Figures M2.2 and M2.3), a template of the scaffolding process when writing generalizations (Figure M2.4), a comprehensive list of sample verbs to use when crafting generalizations (Figure M2.5), and a checklist for crafting generalizations (Figure M2.6).

When writing your own generalizations, try using the stem "Students will understand that . . ." to help focus on what students should understand as opposed to what students will know and do in terms of skills. Here is an example of the three-step guide of generalizations for the topic of quadratics:

Level 1 Students will understand that . . .

The quadratic formula is one method to solve a quadratic equation.

Level 2 How or why?

The quadratic formula describes the roots or zeros of the function and helps solve a quadratic equation.

> The Level 2 generalization is the teaching target and where we focus instruction.

Level 3 So what is the significance or effect?

The expression underneath the square root in the quadratic formula, the discriminant, conveys the nature of the roots, which highlight geometrical features.

Figure 3.7 summarizes the scaffolding process when writing quality generalizations.

FIGURE 3.7: SCAFFOLDING TEMPLATE

Level 1	Students will understand that . . .
Level 2	How or why? (Choose which question is most appropriate)
Level 3	So what? What is the significance or effect?

Writing quality generalizations takes practice and is a skill that can be mastered. The teaching target is the Level 2 generalization; use Level 3 for extension or clarification.

We use Level 1 generalizations to show how to avoid "no no" verbs. Our goal is to write generalizations at Level 2 and skip Level 1.

Consider the following generalization:

Level 1: The coefficients of a quadratic function in a connected equation affect the quadratic formula.

Level 1 generalizations are weak and do not help learners to fully understand what and why they are learning. They often include "no no" verbs such as *is, are, have, impact, affect,* and *influence.* To help learners understand what they are learning, ask the questions "How?" or "Why?" to help scaffold a generalization to the next level. Scaffolding gives learners greater clarity and conceptual specificity, which results in depth of understanding.

The above Level 1 generalization becomes a Level 2 generalization:

Level 2: The quadratic formula utilizes the coefficients of a quadratic function in a connected equation and describes the roots of a quadratic function.

Level 1 generalizations should be deleted once the scaffolding process offers a stronger Level 2 generalization.

Level 3 generalizations address the question, "So what is the significance or effect?" To scaffold further to extend our students' learning, we ask the question, "So what?" which explains the purpose of the generalization. It is important to note, however, that in mathematics the critical understanding is usually expressed in Level 2. Level 3 can be used to express the real-world value of mathematics or to mathematically extend the Level 2 understanding with greater conceptual specificity, if appropriate.

Level 3: Different methods of solving quadratic equations distinguish the roots of the quadratic equation, which contribute to providing a graphical representation of the function, which, in turn, describes real-life problems.

or

Level 3: The expression underneath the square root in the quadratic formula, the discriminant, conveys the nature of the roots, which highlight geometrical features.

Figures 3.8 and 3.9 show some more examples of generalizations for the topic of sequences and series and logarithms.

FIGURE 3.8: SCAFFOLDING GENERALIZATION FOR SEQUENCES AND SERIES

Level 1	Students will understand that . . . *Arithmetic and geometric sequences and series are structured by a set of formulae and definitions.* This generalization contains a "passive voice" verb, which weakens the sentence. Flipping the sentence emphasizes the strong verb for a Level 2 generalization.
Level 2	How or why? *A set of formulae and definitions structure arithmetic and geometric sequences.* *Whether arithmetic and geometric sequences and series share a common difference or a common ratio distinguish one from another.*
Level 3	So what? What is the significance or effect? *Arithmetic and geometric sequences and series describe patterns in numbers and supply algebraic tools that help to solve real-life situations.*

Teaching target is Level 2 generalizations

FIGURE 3.9: SCAFFOLDING A PROCESS GENERALIZATION

Level 1	Students will understand that . . . *Logarithm laws affect logarithmic expressions.*
Level 2	How or why? *Logarithm laws reduce logarithmic expressions by utilizing the inverse process of exponential functions, which represent continuous compounded growth.*
Level 3	So what? What is the significance or effect? *Logarithm laws provide a means of changing multiplicative processes into additive processes, and this can provide the means to find inverses of exponential functions (continuous compounded growth), which have real-world applications.*

In a particular unit of work, not all generalizations will move toward Level 3, as Level 2 generalizations provide strong, clear statements of conceptual understandings. Ensure that your generalizations include the key concepts of the unit. Remember the focus of your teaching is on Level 2 generalizations.

The companion website provides a comprehensive list of sample verbs to use when crafting generalizations (Figure M2.5).

How Do We Draw Out Conceptual Understandings From Our Students?

Effective mathematics learning arises out of guiding students to particular principles or generalizations through inductive inquiry.

Key to inductive teaching is providing students with specific examples from which to draw generalizations. Traditional curriculum and instruction have focused on deductive approaches: telling students the generalization and then asking students to study specific examples. Concept-based curricula emphasize inductive approaches guiding students, through inquiry, to discover generalizations and principles for themselves.

Figure 3.10 shows an inductive, inquiry-led activity guiding students to understand the following generalization:

The expression underneath the square root in the quadratic formula, the discriminant, conveys the nature of the roots, which highlight geometrical features of the quadratic function.

When using inductive approaches, students need to be given opportunities to form conceptual understandings from their learning experiences by writing in mathematics. The learning experiences should involve students looking at specific examples and seeking patterns in order to devise and write generalizations.

In Figure 3.10, the prompt "Explain in your own words the significance of the expression underneath the square root sign in the quadratic formula and how this

FIGURE 3.10: AN EXAMPLE OF INDUCTIVE INQUIRY TO DRAW A GENERALIZATION

Roots of Quadratic Equations

Look at the four examples of quadratic equations and solve them using the quadratic formula.

1. $3x^2 - x - 1 = 0$
2. $2x^2 - 3x - 5 = 0$
3. $4x^2 - 12x + 9 = 0$
4. $x^2 - x + 1 = 0$

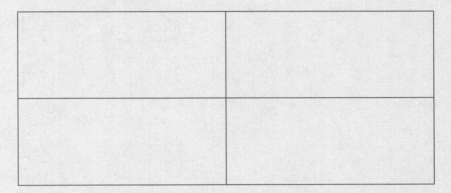

What does solving the above quadratic equations tell you about the associated quadratic functions and their graphs? What is the special name for these values?

What do you notice about what is underneath the square root sign in the four examples above?

Find out what the *"bit"* underneath the square root is called and include an explanation.

Draw a sketch of these four parabolas.

1. $y = 3x^2 - x - 1$
2. $y = 2x^2 - 3x - 5$
3. $y = 4x^2 - 12x + 9$
4. $y = x^2 - x + 1$

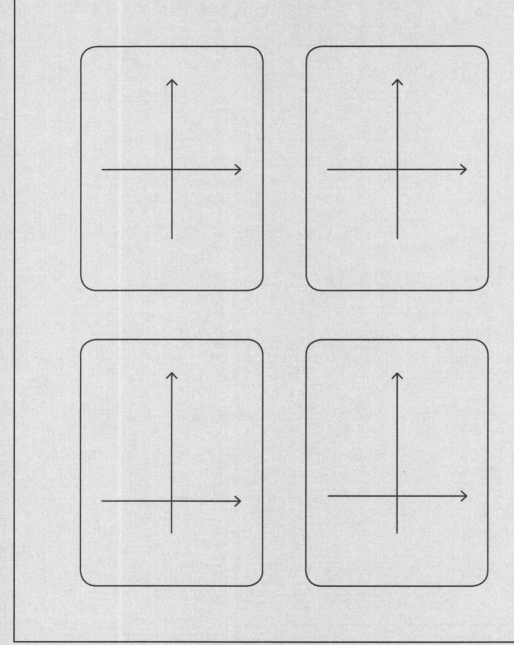

Complete this table:

Value of the _____	Nature of the Roots	Graphical Sketch
$b^2 - 4ac > 0$ and a perfect square		
$b^2 - 4ac > 0$ and not a perfect square		
$b^2 - 4ac = 0$		
$b^2 - 4ac < 0$		

Do you understand the concept?

1. For each part, find the value of the _____ and state whether the equation will have 2 roots, 1 repeated root, or no roots, and include a sketch.

(a) $x^2 + 11x - 2 = 0$

(b) $x^2 - 3x + 3 = 0$

(c) $9x^2 + 6x + 1 = 0$

(d) $x^2 + 13x + 36 = 0$

2. If $2x^2 + bx + 50 = 0$ has a repeated root, find the value of b. Include a sketch.

3. Find the range of values k can take for $9x^2 + kx + 4 = 0$ to have 2 real distinct roots.

Explain in your own words the significance of the expression underneath the square root sign in the quadratic formula and how this helps us with the geometrical significance of a parabola.

 For a completed version of Figure 3.10, please visit the companion website.

helps us with the geometrical significance of a parabola" is used to draw the generalization from students. Research shows that encouraging students to write in the math classroom develops the ability to organize, understand, analyze, and reflect on the new learning (Pugalee, 2005; Stonewater, 2002). Writing allows students to communicate how they have processed information and skills and whether they possess the conceptual understandings of a unit. David Sousa (2015) affirms, "Adding this [writing in math] kinesthetic activity engages more brain areas and helps students to organize their thoughts about the concept" (p. 201).

Graphic organizers are another useful tool for helping draw out generalizations. A graphic organizer can give students a visual representation of how the concepts are related, and they will then be able to form generalizations.

Figure 3.11 is an example of a graphic organizer used to draw generalizations about trigonometry.

Students should be given the opportunity to explore or identify generalizations from their program of study. There are several strategies to help students with this process. Figure 3.12 lists ideas for drawing the generalizations from our students.

Learning experiences should be designed to draw out conceptual understandings, using correct mathematical terminology. If asked about the purpose of the unit, students need to be able to communicate the essence of the desired conceptual understanding verbally and nonverbally.

At the secondary school level, a year-long course may have five or six instructional units. Each unit may have six to eight generalizations to encourage guided inquiry, depending on the length of the unit.

It is important when identifying your conceptual understandings that thought goes into the planning of the entire unit and the concepts involved by utilizing a unit web, as suggested by Erickson and Lanning (2014). A unit webbing tool provides a strategy for including generalizations, which will be discussed in the next chapter. The next chapter will also discuss how to plan a unit of work, including guiding and essential questions to support inquiry leading to deep, conceptual understanding.

FIGURE 3.11: AN EXAMPLE OF A GRAPHIC ORGANIZER TO DRAW GENERALIZATIONS FROM STUDENTS FOR TRIGONOMETRY

What is the big idea?

Using the two strands—right-angled and non-right-angled triangles—write down micro concepts (ideas) in each strand. For each strand, write a generalization (statement) that connects the concepts and captures the most important idea.

What is the big idea?

Using the two strands—right-angled and non-right-angled triangles—write down micro concepts (ideas) in each strand. For each strand, write a generalization (statement) that connects the concepts and captures the most important idea.

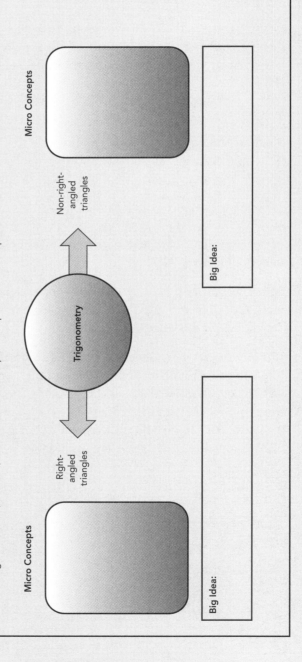

Strategy	Example
Give the concepts and ask students to connect these concepts in a statement that reflects the purpose of their unit of study.	Write the following concepts in a related sentence. This statement should demonstrate your understanding of how these concepts are connected: zeros of a function, roots of an equation, x-intercepts, quadratic formula *or* parallel lines, gradient, transversal, angles
At the end of a lesson or unit of work ask students to complete the sentence, "I understand that . . ."	I understand that solving a quadratic equation by using the quadratic formula graphically displays the x-intercepts of a quadratic function.
Give students a choice of a few statements and ask students to choose one that explains the big idea of the lesson or unit and to justify why.	The quadratic formula is $$x = \frac{-b \pm \sqrt{b^2 - 4ac}}{2a}.$$ Solving a quadratic equation by using the quadratic formula graphically displays the x-intercepts of a quadratic function. The quadratic formula tells us what x is equal to and allows us to solve quadratic equations.
After a lesson or unit of study, ask students to create a concept map, which includes the concepts they have learned and how they are connected, with statements explaining them.	With the word *quadratics* at the center, create a concept map using the ideas you have learned and include statements to explain how they are connected.
Give students a cloze activity. Hint Jar: quadratic equation quadratic formula quadratic function x-intercepts	Solving a _____ _____ by using the _____ _____ graphically displays the _____ of a _____.
Ask students to devise a headline that summarizes the key ideas (concepts) of a unit.	"SOHCAHTOA helps us to find angles and lengths of sides in right-angled triangles."

Chapter Summary

This chapter helps educators develop quality generalizations to foster students' conceptual understanding. Generalizations are statements of conceptual understanding and represent our goals for learning. They are also known as essential understandings, enduring understandings, or the big ideas in a unit of work.

There are two types of enduring understandings (generalizations): overarching and topical. This chapter discusses how to craft topical generalizations that are more specific to a unit of work and focus on particular insights we want students to acquire. In mathematics, generalizations describe the relationships between important concepts. Principles include mathematical theorems, which are cornerstone truths.

Lynn Erickson provides a three-step scaffolding process when writing generalizations and suggests avoiding certain overused verbs to create stronger statements of understanding. Scaffolding generalizations supports the process of crafting high-quality statements of conceptual understanding. Instruction focuses on the Level 2 generalizations, and Level 3 is provided if more clarification or extension is appropriate. Units of work for a concept-based curriculum mainly focus on Level 2 generalizations, with some Level 3 to support the inductive teaching process.

There are several strategies that can be used to draw generalizations from learners. Examples include using the structured or guided inquiry task or utilizing a graphic organizer that asks for the main ideas and concepts of a unit. At the secondary school level, a year-long course may have five or six instructional units. Each unit may have six to eight generalizations to encourage guided inquiry.

The next chapter looks at the power of the unit web and planner when designing concept-based curriculum and instruction.

Discussion Questions

1. What is the difference between a principle and a generalization in mathematics?
2. Why do we wish our students to understand principles and generalizations in mathematics?
3. What is the difference between Level 1, 2, and 3 generalizations?
4. What opportunities do you provide for your students to demonstrate and communicate their conceptual understanding?
5. How will you develop your skills at crafting generalizations when planning your units of work?

CHAPTER 4

How Do I Plan Units of Work for a Concept-Based Curriculum?

A **unit web** is an essential tool for planning a concept-based unit of work. A unit web contains critical content topics and concepts and gives an overview of the depth and breadth of the unit of instruction. Erickson (2007) has said the more complete the planning web, the stronger the conceptual understanding. When planning a unit of work for mathematics, it is important to include a strand titled "Concepts in Mathematical Processes," which represents the process generalizations specific to the unit. The other strands around the unit web represent the knowledge generalizations.

Collaborative unit planning empowers teachers to take control of the learning in lessons and have an input into designing curriculum. Concept-based curriculum and instruction is a model that promotes thinking and can overlay any standards or prescribed curriculum.

Teachers feel ownership of the classroom and lessons again. No longer are they just following a textbook but they are allowed to be more creative in the design of their lessons and therefore more in control of learning. Concept-based approaches are so different to how teachers learned how to teach or how they were taught so teachers are more involved, enthusiastic and able to focus learning on the underlying concepts

Jan Frenis, Math Supervisor
Waterbury Public Schools, Connecticut

FIGURE 4.1: UNIT WEB FOR FUNCTIONS

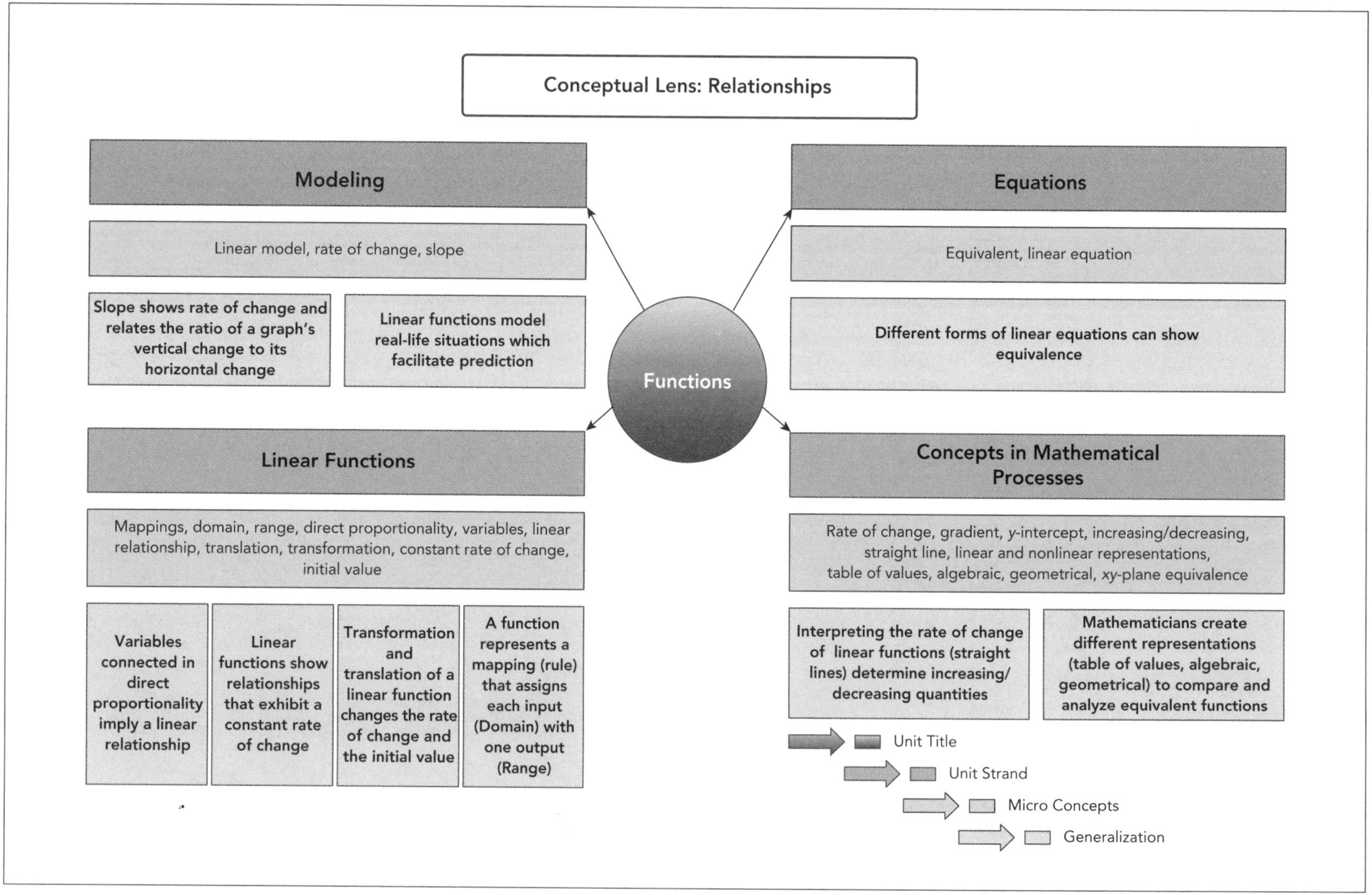

Conceptual Lens: Relationships

Functions

Modeling

Linear model, rate of change, slope

Slope shows rate of change and relates the ratio of a graph's vertical change to its horizontal change

Linear functions model real-life situations which facilitate prediction

Linear Functions

Mappings, domain, range, direct proportionality, variables, linear relationship, translation, transformation, constant rate of change, initial value

Variables connected in direct proportionality imply a linear relationship

Linear functions show relationships that exhibit a constant rate of change

Transformation and translation of a linear function changes the rate of change and the initial value

A function represents a mapping (rule) that assigns each input (Domain) with one output (Range)

Equations

Equivalent, linear equation

Different forms of linear equations can show equivalence

Concepts in Mathematical Processes

Rate of change, gradient, y-intercept, increasing/decreasing, straight line, linear and nonlinear representations, table of values, algebraic, geometrical, xy-plane equivalence

Interpreting the rate of change of linear functions (straight lines) determine increasing/decreasing quantities

Mathematicians create different representations (table of values, algebraic, geometrical) to compare and analyze equivalent functions

Unit Title

Unit Strand

Micro Concepts

Generalization

Unit Webs

Let us take a look at a specific example of a unit web for the topic "functions." Figure 4.1 shows the functions unit web, which include the knowledge strands and the concepts in mathematical processes strand.

For a generic unit web template and a template for a Concept-Based Unit Planner, please visit Figures M4.1 and M4.2 on the companion website.

What features do you notice in the unit web in Figure 4.1? Notice the meso concept of "functions" is in the center of the unit web and connects the strands modeling, equations, linear functions, and concepts in mathematical processes. Strands around this web are the major categories for functions. Modeling, equations, and linear functions are the knowledge strands in the web, while the concepts in mathematical processes represents the processes strand.

The conceptual lens of "relationships" focuses the unit work and provides conceptual depth by stimulating synergistic thinking. Erickson and Lanning (2014) confirm, "The lens ensures synergistic thinking as the student processes factual knowledge through the focus concept" (p. 99).

Erickson explains the purpose of a conceptual lens is to create a synergy between the factual and conceptual levels of thinking (**synergistic thinking**). A conceptual lens focuses work in the instructional unit and facilitates the transfer of understanding through time, across cultures, and across situations. In the International Baccalaureate (IB) Middle Years Program (MYP), the mathematical conceptual lenses are referred to as "key concepts" and they include form, logic, and relationships.

Conceptual lenses such as change, patterns, structure or rules, randomness, or uncertainty help focus the concepts in mathematics; they converge ideas and suggest a focus for transferable ideas. Figure 4.2 summarizes wonderful examples of conceptual lenses for mathematics.

FIGURE 4.2: EXAMPLES OF CONCEPTUAL LENSES FOR MATHEMATICS

Randomness	Structure	Patterns	Change	Uncertainty
Form	Relationships	Systems	Order	Space

Generalizations (also known as essential, enduring understandings) for each strand incorporate the micro concepts and are represented by the blue boxes in Figure 4.1.

Figure 4.3 shows an example of a generic unit web–planning tool for mathematics.

FIGURE 4.3: GENERIC MATHEMATICS UNIT WEB TEMPLATE

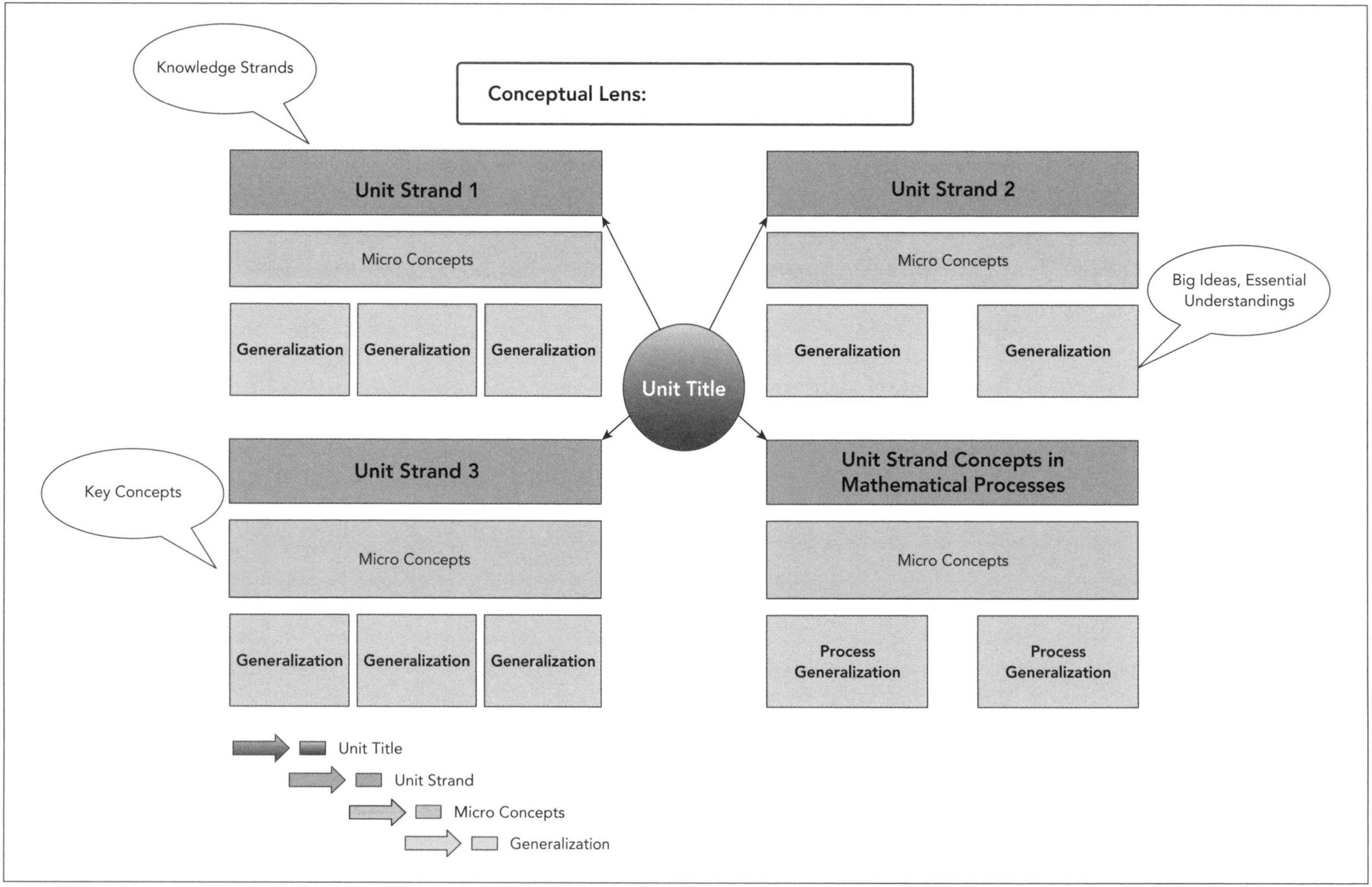

Unit Planning

Understanding by Design (Wiggins & McTighe, 2006a) provides a framework for planning a unit of work, starting with the aims and goals of that unit. The authors have developed three stages of backward design:

1. Identify your results
2. Determine acceptable evidence
3. Plan learning experiences and instruction

This is compatible with Lynn Erickson and Lois Lanning's (2014) 11-step guide to planning a unit of work, shown in Figure 4.4.

FIGURE 4.4: STEP-BY-STEP UNIT PLANNING CHECKLIST

Step	Description
1. Create the unit title.	The unit title can be engaging for students but needs to clearly indicate the content focus.
2. Identify the conceptual lens.	The conceptual lens is a concept that provides focus and depth to the study and ensures synergistic (factual/conceptual) thinking.
3. Identify the unit strands.	The strands will be major headings, which break the unit title into manageable parts for intradisciplinary units. Strands are placed in a web around the unit title. Math units contain a particular strand called "concepts in mathematical processes," which represent the process generalizations.
4. Web out the unit's topics and concepts under the strands.	After brainstorming, underline the concepts under each strand so they can be easily accessed in the next step.
5. Write the generalizations that you expect students to derive from the unit study.	You will craft one or two generalizations using the conceptual lens and one or two generalizations for *each* of the strands. These are what we want our students to understand from their unit of study. Sometimes a generalization will address one or more strands (especially in a process discipline). A unit of study may have five to eight generalizations, depending on the grade level and length.
6. Brainstorm the guiding questions.	Guiding questions facilitate student thinking about the generalizations. Guiding questions should be categorized by type (factual, conceptual, debatable). Each generalization needs a mixed set of three to five factual and conceptual questions developed during the planning process and two or three debatable questions for the unit as a whole.

(Continued)

Step	Description
7. Identify the critical content.	The critical content is the factual knowledge required for grounding the generalizations, deepening knowledge of the unit topic, and defining what students may need to know about processes/skills.
8. Identify the key skills.	Processes are drawn verbatim from academic standards or national curricula. *Processes* transfer across applications and are not tied to specific topics until they appear in the learning experiences. *Skills* in math are smaller strategies that do relate to a specific topic. This entails what we want our students to be able to do in the unit.
9. Write the common, culminating assessment and scoring guide/rubric.	The culminating assessment reveals student understanding of important generalizations, their knowledge of critical content, and key skills. Develop a scoring guide, or rubric, with specific criteria for evaluating student work on the task. Be certain to address the conceptual understanding in the rubric criteria.
10. Roll out suggested learning experiences.	Learning experiences ensure students are prepared for the expectations of the culminating assessment and reflect what students should understand, know, and be able to do by the end of the unit. Learning experiences are meaningful and authentic. Included in this section are suggestions for pacing, other assessment, differentiation strategies, and unit resources.
11. Write the unit overview.	The unit overview is written to read to the students as a hook to grab their interest and attention and to introduce them to the study.

Adapted from Erickson and Lanning (2014). *Transitioning to Concept-Based Curriculum and Instruction,* Corwin Press Publishers, Thousand Oaks, CA.

Both frameworks emphasize the need for teachers to plan with the conceptual understandings at the forefront and not just as accidental add-ons at the end.

When planning units of work, collaboration is the key. Not only will the experience and expertise of your team be utilized, you will also encourage buy in and enthusiasm if everyone is given the opportunity to contribute.

Guiding Questions

Guiding questions are an important component of unit planning once generalizations have been crafted. For examples of guiding questions and a template for writing guiding questions, visit Figures M3.1–M3.3 on the companion website. Lynn Erickson's (2007) work explains three types of guiding questions: factual, conceptual, and provocative/debatable. The purpose of these three types is to help students to connect the factual support with the conceptual understandings. Each generalization in a unit should have a mixture of three to five factual and conceptual questions. A unit may have two to three provocative/debatable questions. The theory of knowledge is a core component of the IB diploma program and addresses the questions of how we know. There are many suggestions in the math guides that can provide debatable questions for a unit. Figure 4.5 shows some examples of questions adapted from the IB Higher Level Math guide, 2014 syllabus:

FIGURE 4.5: EXAMPLES OF DEBATABLE/PROVOCATIVE QUESTIONS

- Were logarithms an invention or discovery?
- Do proofs provide us with completely certain knowledge?
- What determines the validity of a proof?
- Do the words "imaginary numbers" and "complex numbers" make the concepts more difficult than if they had different names?
- Was the complex plane already there before it was used to represent complex numbers geometrically?
- Does studying the graph of a function contain the same level of mathematical rigor as studying the function algebraically (analytically)?

Wiggins and McTighe (2013) list four types of questions:

1. Questions that hook: These questions are in the teacher's toolbox to engage students and capture their interest and attention. They may be used at the beginning of a unit or lesson. Here some examples of "questions that hook" to help introduce calculus and the concept of a gradient function:

    Who has ever climbed Mount Everest?

    Who has ever gone surfing?

    Who loves to go hiking on weekends?

2. Questions that lead: These are closed and offer one answer:

    Can you find the derivative of $y = (x^2 - 4)e^x$?

3. Questions that guide: These are common in inquiry tasks and thought-provoking for students. Here are some examples of questions that guide when introducing calculus:

    Can you predict the next answer?

    Do you notice a pattern?

4. Essential questions: These tend to raise more questions and require deeper understanding, so they often involve a revision of answers and different possible answers.

    When does $f((g(x)) = g(f(x))$?

Planning a Unit of Work for Functions

Figure 4.6 is an example of a unit planner for the meso concept "functions." The unit overview is the introduction to the unit of study and is normally written last in

FIGURE 4.6: UNIT PLANNER FOR FUNCTIONS

Unit Title: Functions	Conceptual Lens: Relationships	Time Allocation: 4 weeks / Grade/Year: Grade 8/Year 9
Unit overview: In different countries there are different units for measuring temperature, distance, and even the weights of objects. What mathematical models can we use to convert these measures easily?	Concepts in the Unit: linear function, mapping, domain, range, direct proportionality, linear relationship, variables, translation, transformation, constant rate of change, modeling, linear model, rate of change, initial value, slope, equations, equivalent, linear equation, gradient, y-intercept, increasing/decreasing, straight line, linear and nonlinear, representations (table of values, algebraic, geometrical), xy-plane, equivalence	

What we would like students to know	What we would like students to understand	What we would like students to be able to do
Define, evaluate, and compare functions.	1. Variables connected in direct proportionality imply a linear relationship.	Use functions to model relationships between quantities.
1. Know that a function is a rule that assigns to each input exactly one output. The graph of a function is the set of ordered pairs consisting of an input and the corresponding output.	2. A function represents a mapping (rule) that assigns each input (domain) with one output (range).	Process: Creating representations
	3. Linear functions show relationships that exhibit a constant rate of change.	1. Construct a function to model a linear relationship between two quantities.
2. Know that functions can be represented in different ways (algebraically, graphically, numerically in tables, or by verbal descriptions). For example, given a linear function represented by a table of values and a linear function represented by an algebraic expression, determine which function has the greater rate of change.	4. Transformation and translation of a linear function changes the rate of change and the initial value.	Process: Creating representations and making connections
	5. Linear functions model real-life situations, which facilitate prediction.	2. Determine the rate of change and initial value of the function from a description of a relationship or from two (x, y) values, including reading them from a table or from a graph.
		Process: Creating representations and making connections

What we would like students to know	What we would like students to understand	What we would like students to be able to do
3. Know the equation $y = mx + b$ as defining a linear function whose graph is a straight line; give examples of functions that are not linear. For example, the function $A = s^2$, giving the area of a square as a function of its side length, is not linear because its graph contains the points (1, 1), (2, 4), and (3, 9), which are not on a straight line.	6. Slope shows rate of change and relates the ratio of a graph's vertical change to its horizontal change. 7. Different forms of linear equations show equivalence. 8. Interpreting the rate of change of linear functions (straight lines) determine increasing/decreasing quantities. 9. Mathematicians create different representations—table of values, algebraic, geometrical—to compare and analyze equivalent functions.	3. Interpret the rate of change and initial value of a linear function in terms of the situation it models and in terms of its graph or a table of values. Process: Communicating and creating representations 4. Describe qualitatively the functional relationship between two quantities by analyzing a graph (e.g., where the function is increasing or decreasing, linear or nonlinear). Process: Creating representations and communicating 5. Sketch a graph that exhibits the qualitative features of a function that has been described verbally.

Generalization	Guiding Questions	
1. Variables connected in direct proportionality imply a linear relationship.	Factual Questions: What is a variable? What is direct proportionality? Conceptual Questions: How do variables represent direct proportionality? How are variables different from parameters?	

(Continued)

FIGURE 4.6: (CONTINUED)

Generalization	Guiding Questions
2. A function represents a mapping (rule) that assigns each input (domain) with one output (range).	Factual Questions: What do *domain* and *range* mean? Conceptual Questions: How does mapping explain the concept of a function? Why is a mapping diagram useful in describing a function?
3. Linear functions show relationships that exhibit a constant rate of change.	Factual Questions: What is a linear function? Conceptual Questions: How do linear functions show a constant rate of change?
4. Transformation and translation of a linear function changes the rate of change and the initial value.	Factual Questions: What are the four types of transformations? Conceptual Questions: How does $y = mx + b$ represent transformation and translation? How do you interpret the initial value of a linear function?
5. Linear functions model real-life situations, which facilitates prediction.	Factual Questions: What is *modeling* in mathematics? Conceptual Questions: How do linear functions model real-life situations? How can we use linear functions for prediction?
6. Slope shows rate of change and relates the ratio of a graph's vertical change to its horizontal change.	Factual Questions: What is the formula for slope? Conceptual Questions: How does the ratio of vertical change to horizontal change relate to slope? How can determining the slope be useful in linear functions?

Generalization	Guiding Questions
7. Different forms of linear equations are equivalent to each other.	Factual Questions: What are the different forms for the equation of a line? Conceptual Questions: How can you express a linear relationship in different ways? Why is it useful to express linear functions in different forms?
8. Interpreting the rate of change of linear functions (straight lines) determines increasing/decreasing quantities.	Factual Questions: What is a decreasing/increasing function? Conceptual Questions: How does interpreting the rate of change of linear functions help determine whether the quantities are increasing or decreasing? Why do we want to know whether a linear function is decreasing or increasing?
9. Mathematicians create different representations—table of values, algebraic, geometrical—to compare and analyze equivalent functions.	Factual Questions: What is a table of values? What is a graph on an xy-plane? Conceptual Questions: How do mathematicians create different representations to compare and analyze equivalent functions?

Debatable Unit Questions
How well does a linear function fit all situations in real life? How reliable are predictions when using models?

the unit planning process. The unit planner includes the critical content of what we want our students to know (factual content), understand (generalizations), and do (the skills specific to that unit; KUDs). The companion website provides an example of a unit planner for geometry. For additional help with KUDs, see Figures M3.4 and M3.5 on the companion website.

The skills in any unit fall under the umbrella of general processes, such as problem solving, reasoning and proof, making connections, communicating, and creating representations. In Figure 4.6, skills such as "construct a function to model a linear relationship between two quantities" refers to the processes "creating representations and making connections" (or the National Council of Teachers of Mathematics Standards for Mathematical Practice: use appropriate tools strategically; look for and make use of structure; look for and express regularity in repeated reasoning [NCTM, 2014]).

The other important components of the unit planner are the guiding questions (factual, conceptual, and debatable). Each question corresponds to a generalization, with the purpose of drawing out conceptual understandings from students.

Figure 4.7 is a weekly planner for the unit on functions and contains suggestions for different learning experiences and how they relate to the conceptual understandings in this unit.

FIGURE 4.7: WEEKLY PLANNER FOR FUNCTIONS

Functions Weekly Planner		
Timeframe	Micro concepts	**Learning Experiences** to support the understanding of the concepts in the unit: generalizations
Week 1	Mappings Domain Range Direct proportionality Variables Linear relationship Translation Transformation	In order to understand that *Variables connected in direct proportionality implies a linear relationship.* Mapping and function machine "Exploring Simple Mappings" http://nrich.maths.org/6951 Coordinates game activity 1: Figure 4.8 Translating lines and understanding $y = mx + b$: http://nrich.maths.org/6539 Surprising Transformation: http://nrich.maths.org/6544 Direct proportionality short problems: http://nrich.maths.org/9335

		Functions Weekly Planner	
Week 2	Representations Table of values Algebraic Geometrical *xy*-plane Equivalence	In order to understand that *Mathematicians create different representations—table of values, algebraic, geometrical—to compare and analyze equivalent functions.* Using the prompt $y - x = 4$, pose questions such as, "Write down what you know/notice": http://www.inquirymaths.com/home/algebra-prompts/y–x–4 This prompt gives two coordinates and questions posed could be What would the gradient and *y*-intercept be? Write down other coordinates to make a shape: http://www.inquirymaths.com/home/algebra-prompts/coordinates-inquiry Diamond Collector — students need to find the equation of three straight lines to collect as many diamonds as possible: http://nrich.maths.org/5725	
Week 3	Modeling Linear model Rate of change Initial value	In order to understand that *Linear functions model real-life situations, which facilitates prediction.* Parallel lines and same rates of change: http://nrich.maths.org/5609 Modeling project Friday 13th: http://nrich.maths.org/610	
Week 4	Rate of change Gradient *y*-intercept Increasing/ decreasing Straight line Linear and nonlinear	In order to understand that *Interpreting the rate of change of linear functions (straight lines) determines increasing/decreasing quantities.* Fill me up: http://nrich.maths.org/7419	

 For a blank template of a concept-based weekly planner, please visit Figure M4.3 on the companion website.

Figure 4.8 is teacher's notes for a learning experience that helps students understand the concepts of the x and y coordinates in the Cartesian plane.

FIGURE 4.8: TEACHER NOTES FOR COORDINATES GAME

Equipment:

Mini whiteboards (or laminated blank paper)
Markers/whiteboard erasers
Linear equations cards
One chair per student

Purpose and Instructions:

The purpose of this activity is to help students understand coordinates in a Cartesian plane and that variables connected in direct proportionality imply a linear relationship. Ask students to sit in a rectangular array. For example, if you have 30 students, arrange the chairs into 6 rows by 5 columns.

Activity 1
Ask student in position bottom left to be the coordinate (0,0). Now ask all students to work out their position and write their coordinates. Do not tell students what their coordinate is.

Ask students to stand up if they are $x = 0$, $x = 1$, then $x = 2$ then $x = 3$ and $x = 4$
Next go through $y = 0, 1, 2, 3, 4$

Now show the card $y = x$ and ask students to stand up if they follow this rule.
Next show the cards $y = 2x, 3x,$ and $4x$.

Now show $y = x + 1$ and ask students to stand up if they follow this rule.
$y = x + 2$, $y = x + 3$

Option: ask students to change positions and ask them to work their coordinates and repeat the same linear equations.

(0,1)				
(0,0)	(1,0)	(2,0)	(3,0)	(4,0)

Activity 2

Ask everyone to erase their coordinate and let the center student be (0,0).

Ask students to stand up if they are $x = 0$, $x = -1$, then $x = -2$
Next, go through $y = 0, -1, -2$

Now show the card $y = -x$ and ask students to stand up if they follow this rule.
Next show the cards $y = -2x, -3x$
Now show $y = -x + 1$ and ask students to stand up if they follow this rule. Then show $y = -x + 2$
Option: Ask students to change positions and ask them to work their coordinates and repeat the same linear equations.

Alternatively, the student in the top right could be (0,0) so that students to the right are positive along the x-axis and students in front are positive along the y-axis.

(−2.0)	(−1,0)	(0,0)	(1,0)	(2,0)

Planning a Unit of Work for Circles

Figure 4.9 is an example of a unit web for the meso concept "circle geometry." The unit (content) strands are properties, theorems, and radians. These strands are all connected to the lens of proportions.

Figure 4.10 is a unit planner, which includes the unit overview and KUDs (what we want our students to know, understand, and do). In Figure 4.10, specific skills are clearly listed for this particular unit. Skills such as "prove all circles are similar" refer to the process "reasoning and proof" (or "reason abstractly and quantitatively" [NCTM, 2014]). Skills fall under the umbrella of the mathematical processes: problem solving, creating representations, communicating, investigating, reasoning and proof, and making connections.

The other important components of the unit planner are the guiding questions (factual, conceptual, and debatable). Each question corresponds to a generalization, with the purpose of drawing out conceptual understandings from students.

Planning a Unit of Work for Calculus: Differentiation and Integration

The macro concept of calculus is a topic whose subconcepts students often find challenging. Calculus consists of two meso concepts: differentiation and integration. In the UK, the concept of differentiation is taught in Years 9 and 10 (Grades 8 and 9 in the United States), which implies that this concept is accessible to younger students if curriculum and instruction is focused on conceptual understandings. The conceptual lens "change" refers to the concepts in differentiation, while "accumulation" helps build conceptual depth in integration.

Figure 4.11 is a unit web on calculus, which includes the knowledge strands derivative, tangent/normal, stationary points, kinematics, area, volume, and concepts in mathematical processes.

Figure 4.12 is a unit planner, which includes the unit overview and all the information of the unit web in a table format. This includes the critical content of the unit—the KUDs. Skills in this unit include "Draw displacement-time, velocity-time graphs" and "Sketch manually and using a GDC a variety of functions." These relate to the process "creating representations" (or "use appropriate tools strategically" [NCTM, 2014]).

The guiding questions are an important component of the unit planner as the factual, conceptual, and debatable questions help draw conceptual understandings from students.

FIGURE 4.9: UNIT WEB FOR CIRCLE GEOMETRY

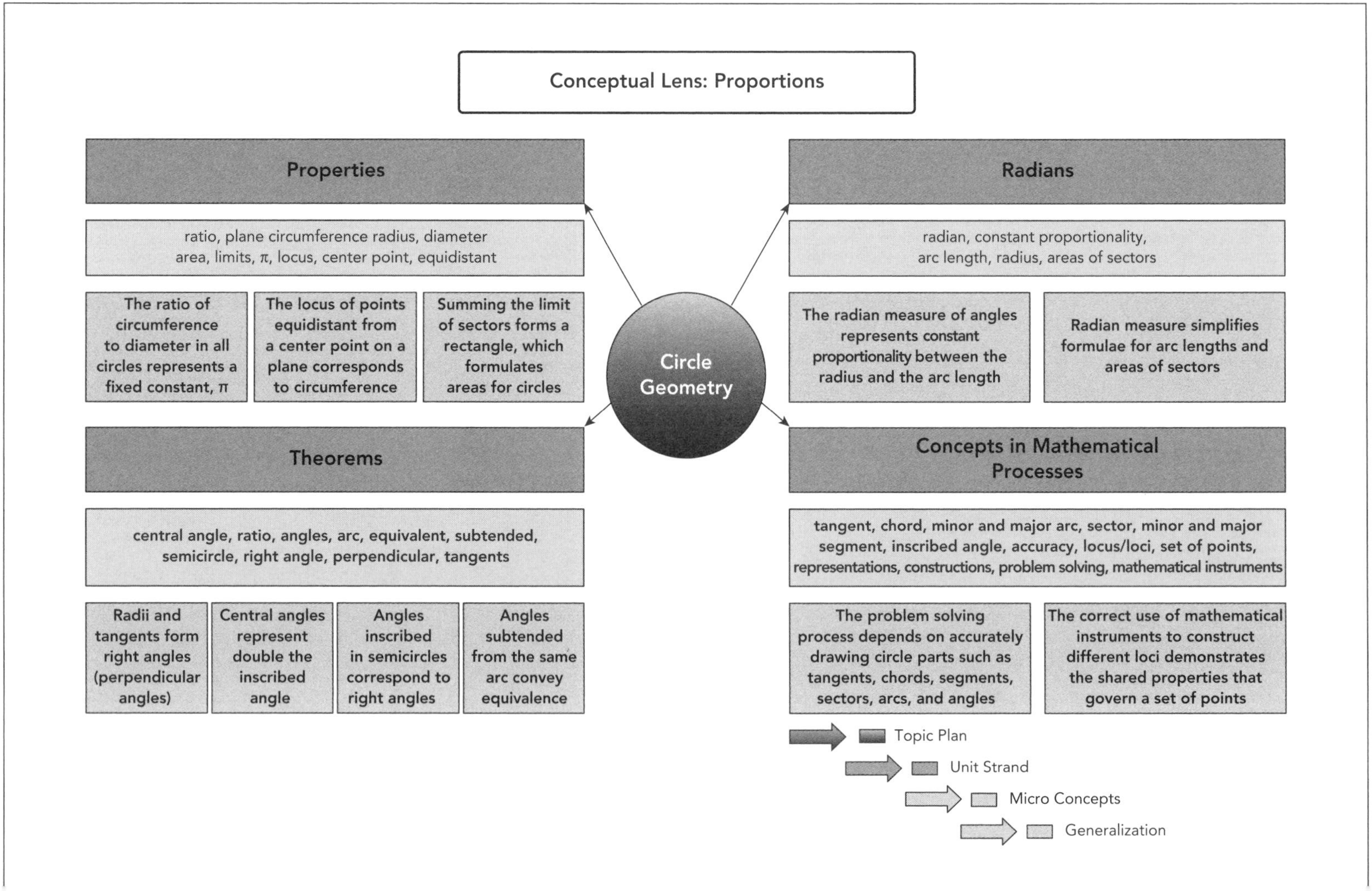

FIGURE 4.10: UNIT PLANNER FOR CIRCLE GEOMETRY

Unit Title: Circle Geometry	Conceptual Lens: Proportions	Time: Allocation: 4–6 weeks	Grade/Year: Grade 8/Year 9
Unit overview: Circles fascinated people for many years. How often do you see a circle in real life? Circles are often used to represent unity and harmony. How would you construct some of the circle designs in ancient artifacts? http://nrich.maths.org/2561		Concepts in the unit: ratio, plane, circumference, radius, diameter, area, limits, π, locus, center point, equidistant, central angle, ratio, angles, arc, equivalence, angle subtended, semicircle, right angle, perpendicular, radian, constant proportionality, arc length, chord, minor and major arc, sector, minor and major segment, inscribed angle, accuracy, locus/loci, set of points Process concepts: representations, constructions, problem solving, mathematical instruments	

What We Would Like Students to Know	What We Would Like Students to Understand		What We Would Like Students to Be Able to Do
1. Know the formula for the area of circumference and area of a circle. 2. Define the radian measure of the angle as the constant of proportionality. 3. Know circle vocabulary: circumference, radius, diameter, chord, minor and major segment, tangent, minor and major arc, locus, right angle, arc length.	1. The ratio of circumference to diameter in all circles represent a fixed constant, π. 2. The locus of points equidistant from a center point on a plane corresponds to circumference. 3. Summing the limit of sectors forms a rectangle, which formulates areas for circles. 4. Angles inscribed in semicircles correspond to right angles. 5. Radii and tangents form right angles (perpendicular angles).		Process: Making connections 1. Convert radians to degrees and vice versa. Process: Reasoning and proof 2. Prove all circles are similar. Process: Reasoning and proof 3. Derive, using similarity, the fact that the length of the arc intercepted by an angle is proportional to the radius, and 4. Derive the formula for the area of a sector.

(Continued)

What We Would Like Students to Know	What We Would Like Students to Understand	What We Would Like Students to Be Able to Do
	6. Central angles represent double the inscribed angle. 7. Angles subtended from the same arc convey equivalence. 8. The radian measure of angles represents constant proportionality between the radius and arc length. 9. Radian measure simplifies formulae for arc lengths and areas of sectors. 10. The problem-solving process depends on accurately drawing circle parts such as tangents, chords, segments, sectors, arcs, and angles. 11. The correct use of mathematical instruments to construct different loci demonstrates the shared properties that govern a set of points.	Process: Creating representations and reasoning and proof 5. Construct the inscribed and circumscribed circles of a triangle, and prove properties of angles for a quadrilateral inscribed in a circle. 6. Construct a tangent line from a point outside a given circle to the circle.

Generalizations	Guiding Questions
1. The ratio of circumference to diameter in all circles represent a fixed constant (π).	Factual Questions: What is the formula to find the circumference and area of a circle? What do the following words mean? circumference, radius, diameter, chord, minor and major segment, tangent, minor and major arc, locus, right angle, arc length Conceptual Questions: How are all circles similar? Why does the ratio of circumference to diameter in all circles represent a fixed constant? How does the ratio of circumference relate to the diameter in all circles?

Generalizations	Guiding Questions
2. The locus of points equidistant from a center point on a plane corresponds to circumference.	Factual Questions: What is the definition of circumference? What does *locus of points* mean? Conceptual Questions: How does the locus of a set of points explain circumference?
3. Summing the limit of sectors forms a rectangle, which formulates areas for circles.	Factual Questions: What is the formula for the area of a circle? Conceptual Questions: How do you find the area of a circle by using the limiting sum? How is a circle similar to a rectangle?
4. Angles inscribed in semi circles correspond to right angles. 5. Radii and tangents form right angles (perpendicular angles). 6. Central angles represent double the inscribed angle. 7. Angles subtended from the same arc convey equivalence.	Factual Questions: What are the circle theorems? Conceptual Questions: How do you identify and describe relationships among inscribed angles, radii, and chords? Include the relationship between central, inscribed, and circumscribed angles; inscribed angles on a diameter are right angles; the radius of a circle is perpendicular to the tangent where the radius intersects the circle

(Continued)

Generalizations	Guiding Questions
8. The radian measure of angles represents constant proportionality between the radius and arc length. 9. Radian measure simplifies formulae for arc lengths and areas of sectors. Arc lengths and areas of sectors intercepted by the angle relate to the radius.	Factual Questions: What is the definition of a radian? Conceptual Questions: Why are radians dimensionless? How is radian measure represented? How does the concept of "constant proportionality" relate to the radian measure of angles?
10. The problem-solving process depends on accurately drawing circle parts such as tangents, chords, segments, sectors, arcs, and angles.	Factual Questions: What is a scale diagram? Conceptual Questions: How do you use constructions to solve problems? Why are constructions useful in the problem-solving process? When is the accurate drawing of circle parts critical for problem solving?
11. The correct use of mathematical instruments to construct different loci demonstrates the shared properties that govern a set of points.	Factual Questions: What are mathematical instruments/applets? Conceptual Questions: How do you use construction/applets to demonstrate shared properties that govern a set of points? Why are applets useful in circle geometry?

Debatable Unit Questions
Which is better for measuring angles: radians or degrees?

FIGURE 4.11: MESO CONCEPT: CALCULUS UNIT WEB

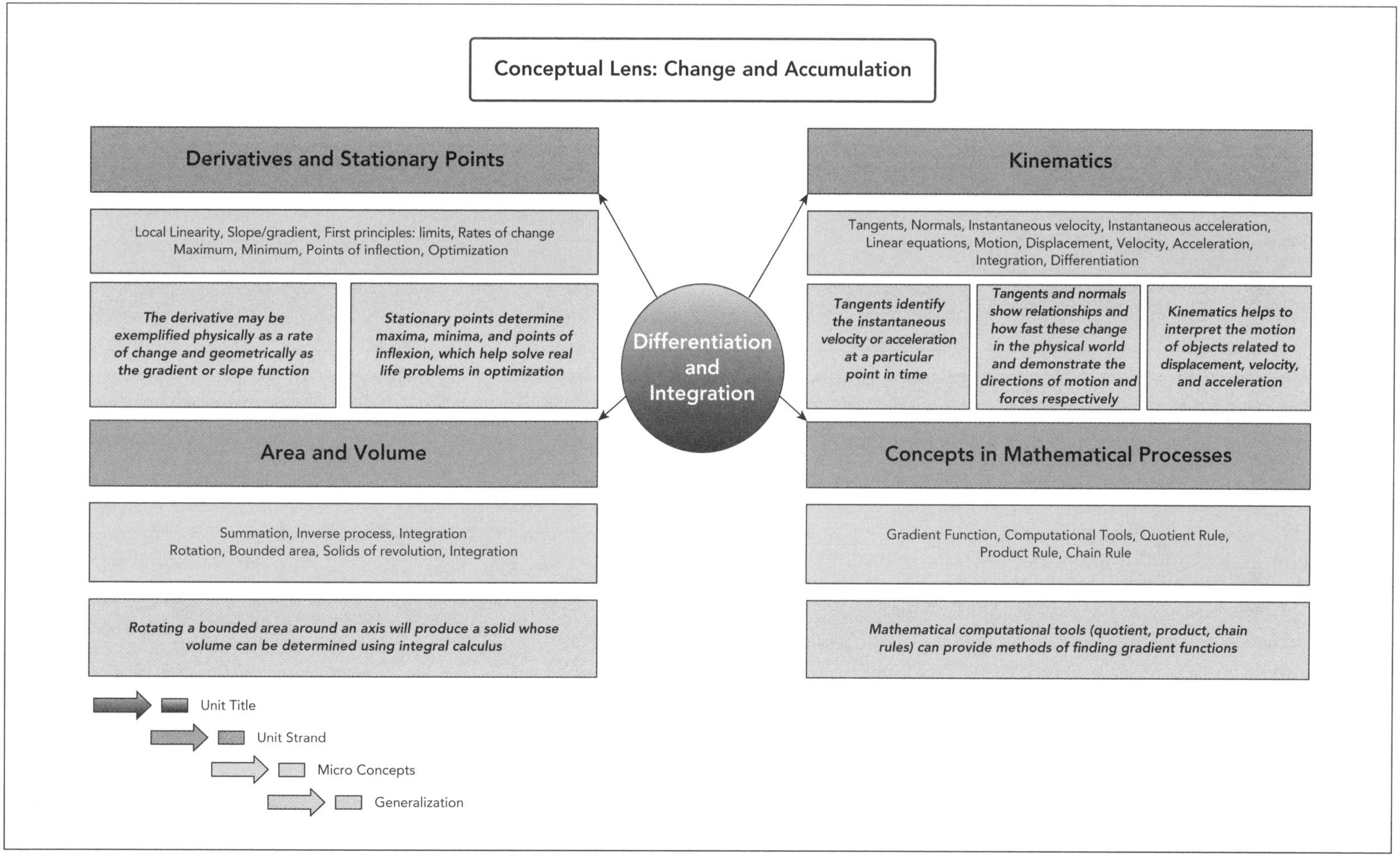

Conceptual Lens: Change and Accumulation

Derivatives and Stationary Points

Local Linearity, Slope/gradient, First principles: limits, Rates of change Maximum, Minimum, Points of inflection, Optimization

The derivative may be exemplified physically as a rate of change and geometrically as the gradient or slope function

Stationary points determine maxima, minima, and points of inflexion, which help solve real life problems in optimization

Area and Volume

Summation, Inverse process, Integration Rotation, Bounded area, Solids of revolution, Integration

Rotating a bounded area around an axis will produce a solid whose volume can be determined using integral calculus

Differentiation and Integration

Kinematics

Tangents, Normals, Instantaneous velocity, Instantaneous acceleration, Motion, Displacement, Velocity, Acceleration, Linear equations, Motion, Differentiation

Tangents identify the instantaneous velocity or acceleration at a particular point in time

Tangents and normals show relationships and how fast these change in the physical world and demonstrate the directions of motion and forces respectively

Kinematics helps to interpret the motion of objects related to displacement, velocity, and acceleration

Concepts in Mathematical Processes

Gradient Function, Computational Tools, Quotient Rule, Product Rule, Chain Rule

Mathematical computational tools (quotient, product, chain rules) can provide methods of finding gradient functions

Unit Title

Unit Strand

Micro Concepts

Generalization

103

FIGURE 4.12: UNIT PLANNER FOR CALCULUS

Unit Title: Calculus	Conceptual Lens: Change and Accumulation	Time Allocation: 10 weeks	Grade/Year: Grade 11/Year 12
Unit overview: Do you know anyone who has climbed Mt Everest? Who loves to go hiking on weekends? We know how to find the slope of straight lines, but how do we find the slope of mountains? Calculus comes from the Latin for "small stone." Differential calculus divides something into small pieces to find out how it changes. Integral calculus joins small pieces together to see how much there is in terms of area and volume.	Concepts in the Unit: derivative, geometrically: gradient/slope, limits, physically: rate of change, function, tangents/normals, instantaneous velocity and acceleration, gradient function, computational tools: chain, product and quotient rules, stationary points, optimization, maxima, minima, points of inflection, differentiation, integration, inverse processes, area, summation, limits, kinematics, displacement, velocity, acceleration, volume, rotating, plane, revolution, bounded areas, three dimensional		

What we would like students to know	What we would like students to understand	What we would like students to be able to do	
1. The definition of gradient/slope, function, derivative 2. Know $y = mx + c$	1. The derivative may be exemplified physically as a rate of change and geometrically as the gradient or slope function. 2. Tangents identify the instantaneous velocity or acceleration at a particular point in time. 3. Tangents and normals show relationships and how fast these change in the physical world. 4. The equation of the tangent and the normal demonstrates the direction of motion and forces, respectively.	Process: Making connections 1. Use appropriate methods for finding the gradient function. Process: Creating representations 2. Draw displacement-time, velocity-time graphs.	

What we would like students to know	What we would like students to understand	What we would like students to be able to do
	5. Mathematical computational tools can provide methods of finding gradient function.	Process: Creating representations
	6. Stationary points determine maxima, minima, and points of inflection, which help solve real-life problems in optimization.	3. Sketch manually and using a GDC a variety of functions.
	7. Differentiation and integration uncover inverse processes of each other.	
	8. The fundamental theorem of calculus presents a method for evaluating a definite integral without having to go back to the definition of taking the limit of a sum of rectangles.	
	9. Kinematics helps to interpret the motion of objects related to displacement, velocity, and acceleration.	
	10. Rotating a plane curve about a line in the same plane with the curve may be looked at as three-dimensional measurement using calculus to find the volume of solids of revolution,	
	or	
	Rotating a bounded area around an axis will produce a solid whose volume can be determined using integral calculus.	

(Continued)

Generalizations	Guiding Questions
1. The derivative may be exemplified physically as a rate of change and geometrically as the gradient or slope function.	Factual Questions: What is the derivative of a function? Conceptual Questions: How do you explain the gradient function? Explain giving an example. How can the derivative be represented? List the various ways.
2. Tangents identify the instantaneous velocity or acceleration at a particular point on time. 3. Tangents and normals show relationships and how fast these change in the physical world. 4. The equation of the tangent and the normal demonstrates the direction of motion and forces respectively.	Factual Questions: What is a tangent? What is a normal? Conceptual Questions: How do tangents describe instantaneous velocity and acceleration? How do tangents and normals relate to real-life situations?
5. Mathematical computational tools can provide methods of finding gradient function.	Factual Questions: What is the chain, product, and quotient rule? Conceptual Questions: How would you explain all the differentiation rules and in which situation each should be employed? How do you determine the most efficient strategy?
6. Stationary points determine maxima, minima, and points of inflection, which help solve real-life problems in optimization.	Factual Questions: What are stationary points of a function? Conceptual Questions: How do you describe stationary points and how does this apply to real-life situations? Why are stationary points given this name?

Generalizations	Guiding Questions
7. Differentiation and integration uncover inverse processes of each other. 8. The fundamental theorem of calculus presents a method for evaluating a definite integral without having to go back to the definition of taking the limit of a sum of rectangles.	Factual Questions: What is the fundamental theorem of calculus? Conceptual Questions: How does the process of summation help when finding the area under a curve? How would you describe the limiting sum?
9. Kinematics helps to interpret the motion of objects related to displacement, velocity, and acceleration.	Factual Questions: What is velocity and acceleration if you are given displacement? Conceptual Questions: How does finding the rate of change help us understand kinematics? How do displacement, velocity, and acceleration relate in terms of a rate of change?
10. Rotating a plane curve about a line in the same plane with the curve may be looked at as three-dimensional measurement using calculus to find the volume of solids of revolution, or Rotating a bounded area around an axis will produce a solid whose volume can be determined using integral calculus.	Factual Questions: What is the formula to find the volume of a solid of revolution? Conceptual Questions: How does integral calculus help us to find volumes of solids of revolution?

Debatable Unit Questions

What value does the knowledge of limits have?

Do limits apply to real-life situations?

Figure 4.13 is a weekly planner, which includes some examples of student learning experiences and the generalization they correspond to. These learning experiences can be supplemented with other materials to reinforce the understanding of the concepts differentiation and integration.

Figures 4.14 to 4.20 illustrate a variety of learning experiences to facilitate the understanding of the topics in calculus. Figure 4.14 is an introductory activity that introduces the idea of gradient function and leads into differentiation from first principles.

Figure 4.15 helps students to understand the idea of increasing and decreasing functions and how to find where functions increase and decrease.

Figure 4.16 leads students to investigate different types of stationary points by looking at the gradient function and the change in sign on either side of the stationary point.

Figure 4.17 is a learning experience that helps students to discover the product rule through an inductive, inquiry-based approach.

Figure 4.18 is a set of problems that relate calculus to the physical world. Students may be given one question each to solve, or groups of four to six students may be given one of the five problem to discuss and present on a poster.

Figure 4.19 contains examples of individual student solutions to the questions posed in Figure 4.17.

Figure 4.20 is a learning experience that leads students to understand the inverse process of differentiation: integration.

Starting with a unit web-planning tool and identifying the content and concepts allows teachers to devise an overall view of the unit in a well-thought-out fashion. Utilizing a unit web allows teachers to determine the unit title, the conceptual lens, the strands to be addressed, the critical micro concepts, and the generalizations in a unit. The unit weekly planner is a wonderful opportunity for teachers to collaborate and share student learning experiences.

FIGURE 4.13: CALCULUS WEEKLY PLANNER

		Calculus Weekly Planner		
Timeframe	Micro concepts	Learning Experiences to support the understanding of the concepts in the unit: generalizations
Week 1	Local linearity Gradients and slopes Function Equations of tangents and normals Increasing, decreasing	In order to understand that: *The derivative may be exemplified physically as a rate of change and geometrically as the gradient or slope function.* Using GDC or graphing software: Gradients and slopes: Figure 4.14. This worksheet is an introduction to gradient functions and how to find the derivative of $y = x^n$ Increasing/decreasing functions: Figure 4.15. This worksheet guides students to understanding what values the gradient function takes when a function is increasing or decreasing.
Week 2	Stationary points: maximum, minimum, and points of inflection	In order to understand that: *Stationary points determine maxima, minima and points of inflection, which help solve real life problems in optimization.* Understanding the concepts of Stationary Points: Figure 4.16. This worksheet will guide students to investigate what happens when the gradient function is zero. Introduction to points of inflection: http://nrich.maths.org/7197
Week 3	Chain, product, and quotient rules	In order to understand that: *Mathematical computational tools can provide methods of finding gradient functions.* Product Rule: Figure 4.17. This is an inductive activity that helps students to generalize and discover the product rule.
Week 4	Applications of calculus to the physical world Optimization Kinematics Displacement Velocity Acceleration	In order to understand that: *Kinematics helps to interpret the motion of objects related to displacement, velocity, and acceleration.* Give one of the five problems to each group to work on. Each group then presents their problem to class with explanations and methodology: Figure 4.18. This worksheet promotes collaboration and helps students to study the real-life applications of calculus.

(Continued)

		Calculus Weekly Planner

Week 5	Fundamental theorem of calculus
	
	Inverse processes
	
	Differentiation
	
	Integration
		In order to understand that:
		
		Differentiation and integration uncover inverse processes of each other.
		
		The fundamental theorem of calculus presents a method for evaluating a definite integral without having to go back to the definition of taking the limit of a sum of rectangles describes.
		
		  1. Integration: Figure 4.19. This worksheet helps students to use the inverse process of differentiation to find the integral of different functions.
		
		  2. What is calculus? How do we work out areas and volumes of irregular shapes?
		
		http://www.brainpop.co.uk/maths/algebra/calculus/
		
		This activity is used as a flipped classroom approach. Ask students to watch the video at home and answer questions on integration in class.

Week 6	Fundamental theorem of calculus	1. Integration Matcher. This worksheet is a collaborative activity where students match answers with questions.
		
		https://nrich.maths.org/6412
		
		  2. Differentiation and integration are inverse processes of each other: Tarsia puzzle. This worksheet is a collaborative activity where students match answers with questions.
		
		http://www.mrbartonmaths.com/jigsaw.htm
		
		  3. Calculus Countdown. This game gives students practice with the concepts of differentiation and integration.
		
		http://nrich.maths.org/6552

Week 7 & 8	Volumes of solids of revolution Rotating Plane Revolution Bounded area Three-dimensional	In order to understand that: *Rotating a plane curve about a line in the same plane with the curve may be looked at as three-dimensional measurement using calculus to find the volume of solids of revolution,* or *Rotating a bounded area around an axis will produce a solid whose volume can be determined using integral calculus.* Brimful from: http://nrich.maths.org/6426/note This problem explores the volume of solid of revolution. The numbers involved are awkward, so student will need to be precise and careful.

FIGURE 4.14: GRADIENTS AND SLOPES

Gradients and Slopes

Questions to think about during this lesson:

We know how to find the gradient of lines, but how do we find the gradient of a curve?

Is there a relationship between a function and the gradient of the function at any particular point?

Why do we want to study gradient functions?

On your GDC or graphing software, graph the following functions and complete the table by finding the gradient at each of the given points. Sketch the gradient function on the axes below.

Part A

$y = x^2$

x	−2	−1	0	1	2	3	4
Gradient							

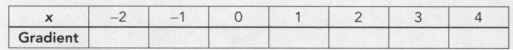

Gradient Function

Is there a relationship between a function and its gradient at any particular point?

Part B

$y = x^3$

x	−3	−2	−1	0	1	2	3
Gradient							

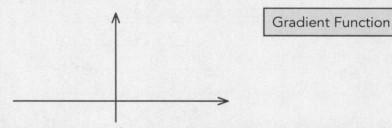

Gradient Function

(Continued)

Is there a relationship between a function and the gradient at any particular point?

Part C

$y = x^4$

x	−3	−2	−1	0	1	2	3
Gradient							

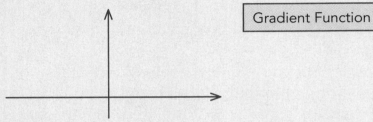

Gradient Function

Is there a relationship between a function and the gradient at any particular point?

Part D

What is the pattern?

Function	Gradient function
$y = x^2$	
$y = x^3$	
$y = x^4$	
$y = x^n$	

Can you explain your rule in words?

The Gradient Function

Find the Gradient of $y = x^2$ at $x = 1$ by completing the table below:

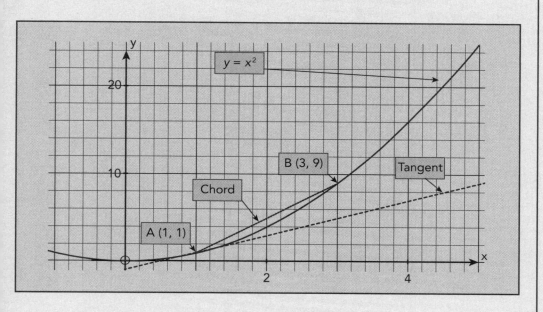

Point A	Point B$_n$ (Remember that point B is on the curve $y = x^2$)	Calculation of Gradient AB$_n$		Gradient AB$_n$
A(1, 1)	B (3, 9)		=	
A(1, 1)	B$_1$ (2.4,)		=	
A(1, 1)	B$_2$ (2,)		=	
A(1, 1)	B$_3$ (1.4,)		=	
A(1, 1)	B$_4$ (1.1,)		=	
A(1, 1)	B$_5$ (1.01,)		=	
A(1, 1)	B$_6$ (1.001,)		=	

The Gradient of $y = x^2$ at the point $x = 1$ is _____

(Continued)

The Gradient Function

Let us look at the curve $y = x^2$ again.
Consider a fixed point **A** and a second point **B** which is very close to **A**.
B is h more in the x direction (by the way, h is very small).
What are the coordinates of B if A is (x, x^2)?

B(   )

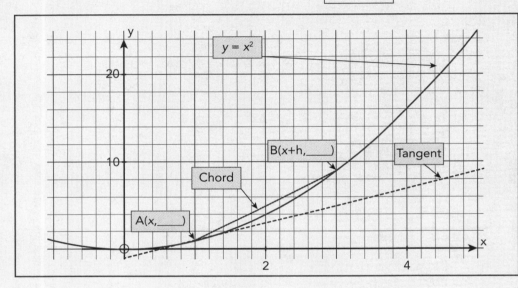

Point A	Point B			
A(x, x^2)	B (x + h,)			
Calculation of Gradient AB				

This method is called "using first principles to find the gradient function."

Since h is very small,	Gradient AB	=	

The gradient for the curve $y = x^2$ at any point on the curve is given by the function:_____

Can you use first principles to find the gradient function for $y = x^3$?

 For a completed version of Figure 4.14, please visit the companion website.

FIGURE 4.15: INCREASING AND DECREASING FUNCTIONS

Increasing and Decreasing Functions

An *increasing* function has a *positive* gradient.
A *decreasing* function has a *negative* gradient.

Look at the gradient of each point of this curve and complete this table. Use your GDC if you need to.

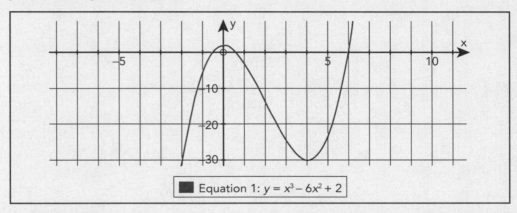

Equation 1: $y = x^3 - 6x^2 + 2$

X	−1	−2	0	1	2	3	4	5	6
Gradient									

What do you notice about the sign of the gradient for $x < 0$? Is the gradient positive or negative? Is this increasing or decreasing?

What do you notice about the sign of the gradient for $0 < x < 4$? Is the gradient positive or negative? Is this increasing or decreasing?

What do you notice about the sign of the gradient for $x > 4$? Is the gradient positive or negative? Is this increasing or decreasing?

Complete this table:

Increasing	Decreasing	Increasing
	$0 < x < 4$	
	$\dfrac{dy}{dx} < 0$	

 For a completed version of Figure 4.15, please visit the companion website.

FIGURE 4.16: STATIONARY POINTS

Stationary Points

A stationary point is a curve that has a gradient of zero or when a curve stops increasing and starts decreasing or vice versa.

Sketch: $f(x) = x^2$	**Sketch:** $f(x) = -x^2$
Where is the stationary point for this curve?	Where is the stationary point for this curve?
What is the sign of the gradient at $x = 1$?	What is the sign of the gradient at $x = 1$?
What is the sign of the gradient at $x = -1$?	What is the sign of the gradient at $x = -1$?
Sketch $f(x) = (x+2)(x-3)(x+1)$ Find out where all the possible stationary points are.	What is a point of inflection? Describe how to find horizontal and nonhorizontal points of inflection. Include a diagram and an example to illustrate your explanation.

 For a completed version of Figure 4.16, please visit the companion website.

FIGURE 4.17: THE PRODUCT RULE

Finding the Derivative of the Product of Two Functions

Part A

1. Find the derivative of this by multiplying out the brackets:

   (a) $y = (2x + 3)(3x - 9)$

   (b) Write down the derivative of $u = 2x + 3$

   (c) Write down the derivative of $v = 3x - 9$

2. Find the derivative of the following by multiplying out the bracket.

   (a) $y = (x^2 + 3)(4x + 1)$

   (b) Write down the derivative of $y = x^2 + 3$

   (c) Write down the derivative of $y = 4x + 1$

3. Think of your own example of finding the derivative of two straightforward functions by multiplying out the brackets. Find the derivative of each of these functions separately.

(Continued)

Part B

Complete this table:

Let the first function be called u(x) and the second function called v(x)

First function $u(x)$	Derivative $\frac{d}{dx}(u(x))$	Second Function $v(x)$	Derivative $\frac{d}{dx}(v(x))$	Derivative of the product $\frac{d}{dx}(u(x) \cdot v(x))$
$2x + 3$		$3x - 9$		
$x^2 + 3$		$4x + 1$		
$u(x)$	$u'(x)$	$v(x)$	$v'(x)$	

The product rule:

$\frac{d}{dx}(u(x) \cdot v(x)) =$

Write down in words a generalization that describes how to find the derivative of the product of two functions:

$$y = u(x) \cdot v(x)$$

Use this generalization to find the derivative of

$$y = (x + 3) \sqrt{x + 4}$$

 For a completed version of Figure 4.17, please visit the companion website.

FIGURE 4.18: REAL-LIFE PROBLEMS FOR CALCULUS

Applications of Calculus to the Physical World

The Open Box Problem

A rectangular sheet of card is 80 cm by 50 cm. A square of side x cm is cut away from each corner of the sheet, which is then folded to make an open box. Let y be the volume of this box in cubic centimeters. Show that $y = 4000x - 260x^2 + 4x^3$. Given that x varies, find the greatest volume of the box.

The Cylindrical Tin

A cylindrical tin closed at both ends is made of thin sheet metal. Find the dimensions of a tin that holds 1000 cubic centimeters and has a total minimum surface area. Give the dimensions to 2 decimal places.

The Spherical Balloon

At time t seconds, the radius, in cm, of an expanding spherical balloon is given by $x = \frac{1}{2} t$. Express the volume V (cubic cm) and the surface area A (square cm) in terms of t. Find the rate of change of V and the rate of change of A with respect to t at $t = 3$ seconds.

The Financial Problem

The profit, $\$y$, generated from the sale of x items of a certain luxury product is given by the formula $y = 600x + 15x^2 - x^3$. Calculate the value of x, which gives a maximum profit, and determine that maximum profit.

Kinematics

The acceleration, a ms^{-2} of a particle moving in a straight line at time t seconds is given by $a(t) = t + 1$. Find the formula for velocity and displacement given that $s = 0$ and $v = 8$ when $t = 0$.

Help Sheet for Applications of Calculus to the Physical World

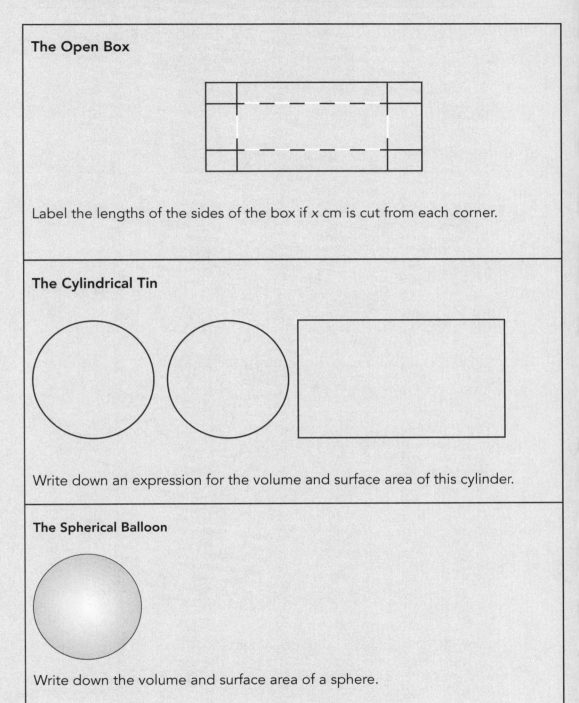

The Open Box

Label the lengths of the sides of the box if x cm is cut from each corner.

The Cylindrical Tin

Write down an expression for the volume and surface area of this cylinder.

The Spherical Balloon

Write down the volume and surface area of a sphere.

(Continued)

FIGURE 4.18: (CONTINUED)

The Financial Problem

How do you find a minimum or maximum stationary point? Use graphing software and sketch the curve $y = 600x + 15x^2 - x^3$. Indicate on your sketch the positions of all stationary points.

Kinematics

Use this relationship:

$$\frac{d}{dx}s(t) = v(t), \frac{d}{dx}v(t) = a(t)$$

FIGURE 4.19: STUDENT SOLUTIONS TO REAL-LIFE PROBLEMS IN CALCULUS

The Open Box Problem Solution by Chun Yu Yiu

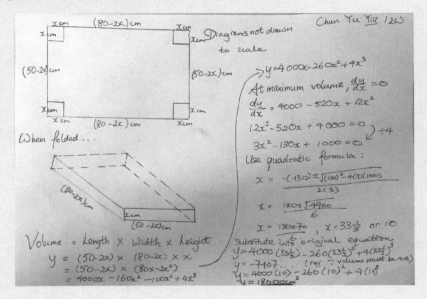

The Cylindrical Tin Solution by Hoi Fung Chow

Volume of cylinder : $V = \pi r^2 h$

Surface area of cylinder : $A = 2\pi r h + 2\pi r^2$

$\pi r^2 h = 1000$

$h = \dfrac{1000}{\pi r^2}$

Substitute h in A

$A = 2\pi r \left(\dfrac{1000}{\pi r^2}\right) + 2\pi r^2$

$= \dfrac{2000}{r} + 2\pi r^2$

$\dfrac{dA}{dr} = -\dfrac{2000}{r^2} + 4\pi r$

Set $\dfrac{dA}{dr} = 0$

$0 = 4\pi r - \dfrac{2000}{r^2}$

$2000 = 4\pi r^3$

$r = \sqrt[3]{\dfrac{500}{\pi}}$

$r = 5.42$

Substitute r in V

$h = \dfrac{1000}{\pi (5.42)^2}$

$h = 10.84$

Hoi Fung Chow

(Continued)

The Spherical Balloon Solution by Cassidy Chan

The Spherical Balloon

$x = \frac{1}{2}t$

$V = \frac{4}{3}\pi r^3$ $\qquad A = 4\pi r^2$

$\quad = \frac{4}{3}\pi(\frac{1}{2}t)^3$ $\qquad = 4\pi(\frac{1}{2}t)^2$

$\quad = \frac{1}{6}\pi t^3$ $\qquad = \pi t^2$

$V' = \frac{1}{2}\pi t^2$ $\qquad A' = 2\pi t$

∴ when $t = 3$, $V' = \frac{1}{2}\pi(3)^2 = \frac{9}{2}\pi$

$\qquad\qquad A' = 2\pi(3) = 6\pi$

The Financial Problem Solution by Cassidy Chan

The Financial Problem

$y = 600x + 15x^2 - x^3$

$\frac{dy}{dx} = -3x^2 + 30x + 600 = 0$

$x = \dfrac{-(30) \pm \sqrt{(30)^2 - 4(-3)(600)}}{2(-3)}$

$\quad = \dfrac{-30 \pm 90}{-6}$

$\quad = -10 \text{ or } 20$

$\frac{d^2y}{dx^2} = -6x + 30$

$x = -10 \Rightarrow \frac{d^2y}{dx^2} = 90$

$x = 20 \Rightarrow \frac{d^2y}{dx^2} = -90$ $\qquad$ ∴ when $x = 20$, it gives a maximum profit

maximum profit:

$\qquad\qquad\qquad\qquad$ ∴ the maximum profit is $10000

$y = 600x + 15x^2 - x^3$

$\quad = 600(20) + 15(20)^2 - (20)^3$

$\quad = 10000$

Kinematics Solution by Rahul Daswani

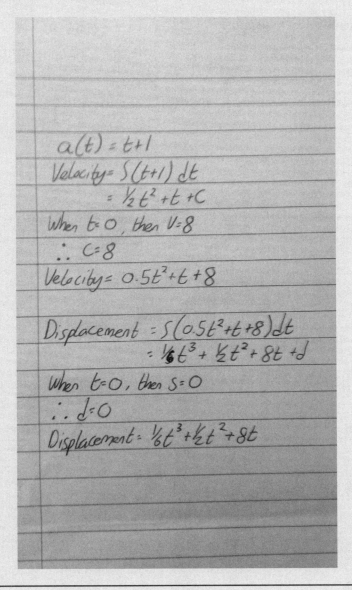

$a(t) = t+1$

Velocity $= \int (t+1)\,dt$

$\qquad = \frac{1}{2} t^2 + t + C$

When $t = 0$, then $V = 8$

$\therefore \; C = 8$

Velocity $= 0.5 t^2 + t + 8$

Displacement $= \int (0.5 t^2 + t + 8)\,dt$

$\qquad = \frac{1}{6} t^3 + \frac{1}{2} t^2 + 8t + d$

When $t = 0$, then $S = 0$

$\therefore \; d = 0$

Displacement $= \frac{1}{6} t^3 + \frac{1}{2} t^2 + 8t$

FIGURE 4.20: INTEGRATION

Integration

Integration is the inverse process of differentiation.

Can you work out the rule to find the integral of any function? Take a guess!

Write down as a function machine the steps to find the derivative of the function $y = x^n$:

$$\frac{d}{dx}(x^n) \longrightarrow \boxed{} \longrightarrow \boxed{} \longrightarrow$$

Using the inverse processes write the above backwards.

$$\int x^n\, dx \longrightarrow \boxed{} \longrightarrow \boxed{} \longrightarrow$$

Complete this table:

$y = f(x)$	$\frac{dy}{dx} = f'(x)$
$y = x^2 + 2$	
$y = x^3 - 4$	
	$f'(x) = 4x^3$
	$f'(x) = x^n$

What happens to a constant when we differentiate?

What do we need to do when we integrate in case there is a constant?

Explain in words how you would find the integral of a function.

 For a completed version of Figure 4.20, please visit the companion website.

Chapter Summary

This chapter discussed the power of a unit web and unit planner. The unit web is an overview of the meso concept under study and includes the conceptual lens, which stimulates synergistic thinking, and the strands, which detail the concepts and content in the unit. The generic unit web template provides a structure of content and concepts for a topic and also provides support to teachers in understanding the components of a concept-based curriculum. Features of a unit web for math may include the meso concept at the center, knowledge strands, and a "concepts in mathematical processes" strand. Generalizations containing micro concepts should be clearly displayed.

Critical components in the unit planners are identifying your KUDs. These refer to what you want students to *know*, what you want them to *understand,* and what you would like them to be able to *do* (skills) from their unit of study. Skills in the unit planner are specific to the unit of work; however, they fall under the umbrella of mathematical processes (or NCTM Standards for Mathematical Practice, 2014): problem solving, reasoning and proof, making connections, communicating, and creating representations. Unit webs and planners were designed for the topic of functions, circles, and calculus in this chapter to provide a model for educators to utilize. These models allow teachers to apply the essential elements when creating and developing their own unit planners.

One of the purposes of a unit web and planner is to promote collaboration and the sharing of ideas within a faculty. Collaboration supports ownership of the material produced and saves time in planning. This chapter guides the reader through the design steps for a concept-based unit in mathematics—from the unit web to the learning experiences through the weekly planner.

The next chapter will look at how to captivate your students by engaging their hearts and minds.

Discussion Questions

1. How does unit planning empower teachers?
2. What are the benefits of a unit-webbing tool?
3. What are the main components for a math unit web?
4. Why do math unit webs include concepts in mathematical processes?
5. How do the three types of essential questions—factual, conceptual, and debatable—support student learning?
6. What is the most effective way to create a unit of work? Describe the process that you would utilize.
7. How will you encourage collaboration when unit planning?

Part III

How Do We Engage Students Through Instructional Practice?

Strategies to Engage and Assess

....

How Do I Captivate Students?

Eight Strategies for Engaging the Hearts and Minds of Students

My goal for every lesson is to captivate all of my students and excite them about learning math. I also believe you are only as good as your last performance (like a Broadway show!), so I strive continually to captivate my students every day in the palms of my hands.

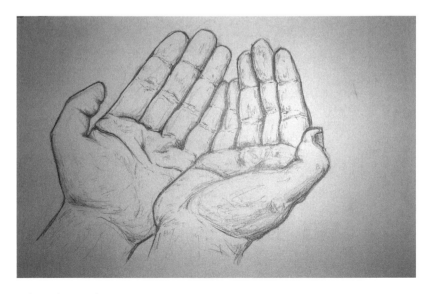

Illustration by Kohana Wilson

Designing an engaging, conceptual, and inquiry-led lesson requires an investment of time and thought to ensure the conceptual understandings emerge during the learning process.

Through my experience as a classroom practitioner, I have compiled my top strategies for designing an engaging, conceptual, and inquiry-led mathematics lesson.

Strategy 1: Create a Social Learning Environment

We are social beings. At work and at school, social interactions are an integral part of each day. Employers value people who collaborate well with others because solving problems and creating new ideas is enhanced when multiple minds and perspectives interact. At school, providing opportunities for students to communicate and construct learning through engaging collaborative tasks not only increases student motivation but also improves conceptual understanding. The collaborative process engages our individual and group intellect, leads to deeper understanding, and offers new insights and creative solutions.

To encourage collaboration, one activity I use is the placemat activity, illustrated by Figure 5.1. The placemat activity can help students to socially construct their understanding of a particular concept. Teachers choose a concept and ask students to write their own description and understanding of the concept on the edges of the placemat. After group discussion, learners are asked to write down an agreed description of the concept in the center.

For example, what is a description of ratio and how do we use this in mathematics? Students have several avenues to take as the concept of ratio encompasses a variety of situations. Students could discuss how ratio is used to compare objects, how it is used in similarity of geometric objects, or they could even talk about ratios in a right-angled triangle.

Michael Swan of the University of Nottingham has developed many strategies to encourage collaboration in the mathematics classroom. In his research paper *Collaborative Learning in Mathematics* (2006a), Swan says, "Traditional, 'transmission' methods in which explanations, examples and exercises dominate do not promote robust, transferable learning that endures over time or that may be used in non-routine situations. They also demotivate students and undermine confidence. In contrast, the model of teaching we have adopted emphasizes the interconnected nature of the subject, and it confronts common conceptual difficulties through discussion." (p. 162)

Swan's (2006a) research found that student-centered, collaborative, and discussion-based approaches were more effective than traditional transmission methods and

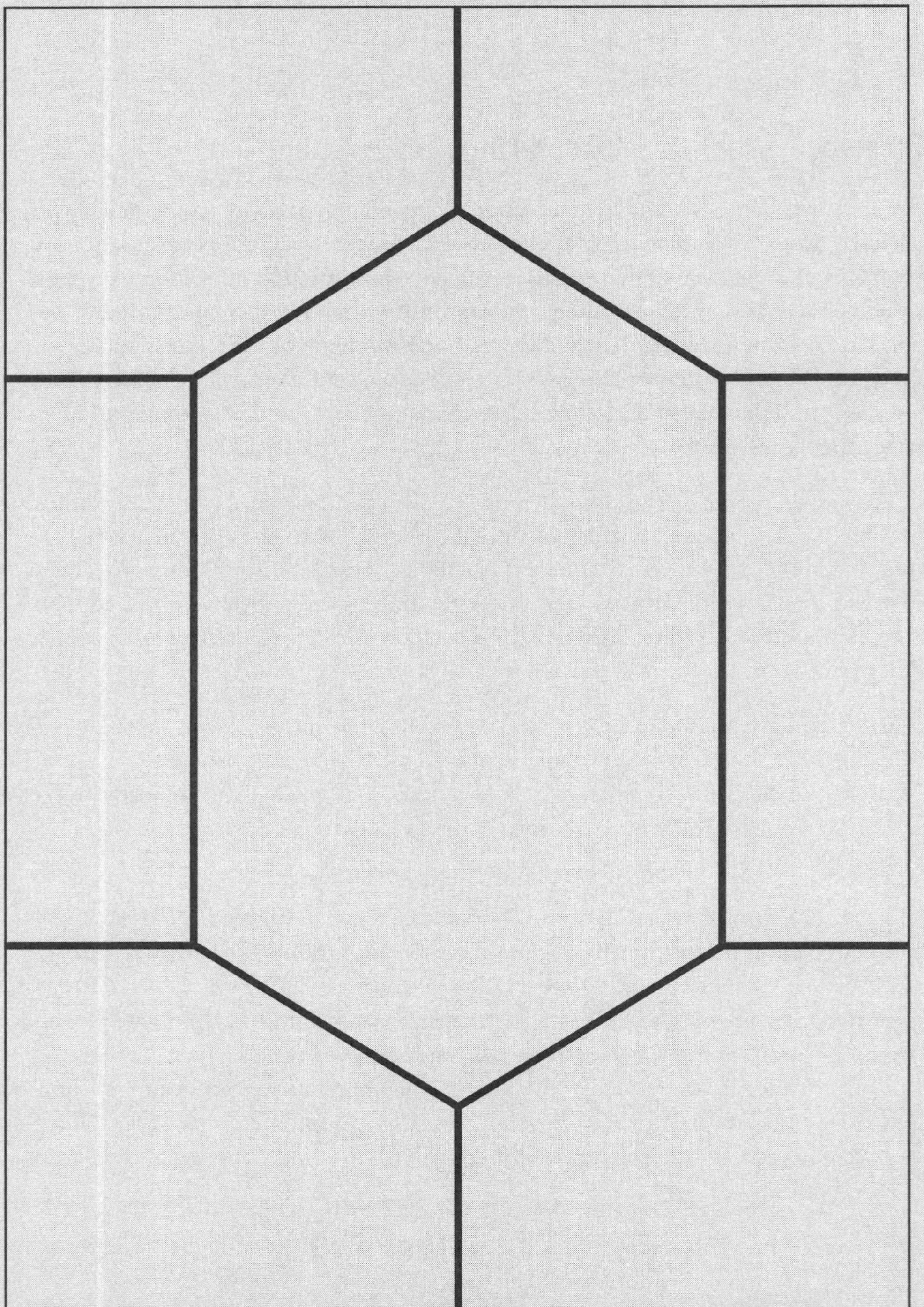

FIGURE 5.1: PLACEMAT ACTIVITY

resulted in improved algebra assessment scores. These improved scores were a result of deeper conceptual understanding of the mathematics developed through collaboration and discussion. There was also a significant increase in pupil motivation and reduction in anxiety around mathematics.

> Swan (2006a) found that collaboration and discussion-based activities in math improve achievement.

Professor Swan (2006a) outlines five teaching strategies that promote collaborative and active learning.

1. Classification of mathematical objects and encouragement of discussion by comparing each other's ideas—for example, classifying different types of quadrilaterals.

2. Matching cards in groups to interpret multiple representations—for example, matching graphs with their characteristics.

3. Evaluating statements through discussion that are always, sometimes, or never true—for example, $4x$ is greater than $x + 5$.

4. Asking students to create problems for each other to solve—this encourages creativity—for example, make up a problem that uses PEMDAS (BODMAS).

5. Comparing methodologies and seeing there are several pathways through a problem—for example, compare methods for multiplying $(x + 3)(x - 1)$.

In a collaborative environment, students participate more in discussions and are given opportunities to give more reasoned explanations. They use mistakes to improve understanding. They learn to listen and reflect on the ideas of others and dialogue in the language of the discipline. They develop mathematics fluency. Working in a collaborative environment aligns with one of the NCTM (2014) Mathematics Teaching Practices: "Facilitate meaningful mathematical discourse. Effective teaching of mathematics facilitates discourse among students to build shared understanding of mathematical ideas by analyzing and comparing student approaches and arguments."

Strategy 2: Provide an Open, Secure Environment to Allow for Mistakes as Part of the Learning Process

Effort and mistakes are important stages of learning. Due to the brain's neuroplasticity when a mistake is made and corrected, new neurons and synapses (connections between neurons) grow and new development of the brain occurs.

> Anyone who has never made a mistake has never tried anything new.
>
> — Often attributed to Albert Einstein
>
> Every time a student makes a mistake in math they grow a synapse.
>
> —Jo Boaler & Carol Dweck

Carol Dweck's (2006) work on growth mindset has explained the brain's enormous capacity to grow. "When you learn new things, these tiny connections in the brain actually multiply and get stronger. The more that you challenge your mind to learn, the more your brain cells grow. Then, things that you once found very hard or even impossible—like speaking a foreign language or doing algebra—seem to become easy. The result is a stronger, smarter brain" (p. 219).

The brain's ability to grow and develop is an extraordinary phenomenon. A case that illustrates this is the story of Cameron Mott. Cameron had the right half of her brain removed due to a rare syndrome. Many were astounded when the left side of her brain took over for the right side and she recovered to function normally after a few years of physiotherapy (Celizic, 2012).

A person with **growth mindset** believes that difficult tasks are opportunities to grow and mistakes are a vital part of this process. Mistakes are a necessary part of learning. Fixed mindset believers tend to flee from challenges and the opportunity to grow. A growth mindset encourages persistence and pursuance of more challenging tasks. In a study of college students, Dweck (2006) measured students' mindsets and concluded, "Once again we found that the students with the growth mindset earned better grades on the course. Even when they did poorly on a particular test, they bounced back on the next ones. When the students with fixed mindset did poorly, they often didn't make a comeback" (p. 61).

Another study by Dweck (2006) looked directly at mathematics achievement levels of two groups. One group was given a series of eight workshops and learned about study skills and growth mindset. The other group was given similar study skills workshops but did not learn about growth mindset. Dweck looked at the effect of the workshops on students' mathematics grades in particular. Students who had received the growth mindset training showed a jump in their grades. Overall, the growth mindset group performed much better than the other students who did not improve their grades despite undergoing an eight-week workshop in study skills.

The important message here is to encourage mistakes in a safe, secure environment. Fight stereotype threats that only certain people can do math, as everybody can do math with hard work, practice, and good facilitation. I know I've succeeded with my students when I hear a mistake being made and corrected and a comment like "YES! I have just grown a new synapse!"

Strategy 3: Use Appropriate Levels of Inquiry and Employ Inductive Approaches to Develop Conceptual Understanding

Concept-based lessons provide a clear path from the facts and processes to the conceptual understandings. Lesson plans should accommodate opportunities for learners to take different lines of inquiry, if appropriate, rather than being rigid or overly prescriptive. Teachers need to provide flexibility during lessons to entertain students' questions and different lines of inquiry. The goal is deeper, conceptual understanding. Provide opportunities for students to discover generalizations—statements of conceptual understanding—by utilizing the inquiry continuum and inductive approaches. Inquiry tasks may be structured, guided, or open according to the confidence of your students and teachers. For any inquiry task, encourage students to start with specific numerical examples that lead to discovering generalizations, as exemplified by an inductive approach. Structured and guided inquiry approaches are explained in Chapter 1. Additionally, the companion website provides exemplars and templates to use when planning inductive inquiry tasks (See Figures M5.1–M5.7).

Ian Bengey, Head of Mathematics, West Island School, Hong Kong encapsulates the importance of inquiry in the following quote: "Inquiry without rigor is careless; rigor without inquiry is thoughtless."

The Sample Student Learning experiences below (Figures 5.2 and 5.3) are provided as examples of structured and guided inquiry tasks, respectively.

Figure 5.4 is an example of a student's explanation after completing the structured inquiry activity. Sean's explanation demonstrates an understanding of the following generalization:

Combinations of the coefficients and the constant term of polynomials determine the sum and product of the roots of polynomials.

From specific examples, he was able to write a generalization about the relationship between the roots of any polynomial and the coefficients of the polynomial.

This task allowed Sean to think in a logical fashion and use his pattern-seeking ability to find the relationship between the coefficients of a polynomial and the roots of a polynomial. He was also encouraged to communicate his understanding through writing mathematics.

Strategy 4: Reduce Whole Class Teacher Talk Time

Teacher talk time (TTT) refers to the time a teacher talks to the whole class. I have noticed more and more in my lessons that every time I talk didactically to my students for more than 15–20 minutes, some students just tune out. I notice these students start to fidget or they pretend to listen. Rather than having students in

●●● Sample Student Learning Experience

FIGURE 5.2: AN EXAMPLE OF A STRUCTURED INQUIRY TASK

Sum of the Roots of Quadratics
Product of the Roots of Quadratics

For the following quadratic functions, find the roots and sketch the function and write down any significant features of these curves. What do the roots of the quadratic function tell us?

1. $y = 3x^2 + 5x$	Sketch
2. $y = x^2 - 5x - 6$	
3. $y = 4x^2 - 4x + 1$	
4. $y = 2x^2 - 13x - 7$	

Complete this table:

Quadratic function	Roots	Sum of Roots	Product of Roots
1. $y = 3x^2 + 5x$			
2. $y = x^2 - 5x - 6$			
3. $y = 4x^2 - 4x + 1$			
4. $y = 2x^2 - 13x - 7$			

From the table, can you make a generalization about the sum and product of roots for a quadratic function?

Quadratic Function	Roots	Sum of roots	Product of Roots
$y = ax^2 + bx + c$	α and β		

In pairs, write down in words the relationship between the sum of the roots and the coefficients for a quadratic equation. Write down in words the relationship between the product of the roots and the coefficients for a quadratic equation. (Generalizations)

(Continued)

The Roots of Cubic Functions

Test to see whether your generalization works for cubic functions. You may use specific examples to try your prediction or use an algebraic proof.

Here are some specific cubic functions for you to try if you are not using the proving and reasoning process. Use your graphing software or a graphical display calculator to find the roots and sketch.

1. $y = (x + 5)(x - 6)(x + 7)$	Sketch
2. $y = 2(x + 3)(x - 3)(x + 1)$	
3. $y = 2(x - 1)(x + 2)(x + 1)$	
4. $y = 2x^3 - 5x^2 - 6x + 4$	

Complete this table:

Cubic function	Roots	Sum of Roots	Product of Roots
1. $y = (x + 5)(x - 6)(x + 7)$			
2. $y = 2(x + 3)(x - 3)(x + 1)$			
3. $y = 2(x - 1)(x + 2)(x + 1)$			
4. $y = 2x^3 - 5x^2 - 6x + 4$			

From the table, can you make a generalization about the sum and product of roots and the coefficients for a cubic function?

Cubic Function	Roots	Sum of Roots	Product of Roots
$y = ax^3 + bx^2 + cx + d$	α, β, and γ		

In pairs, write down in words the relationship between the sum of the roots and the coefficients for a cubic equation. Write down in words the relationship between the product of the roots and the coefficients for a cubic equation. (Generalizations)

(Continued)

The Sum of the Roots and the Product of the Roots of a Polynomial

Quadratic Function	Roots	Sum of Roots	Product of Roots
$y = ax^2 + bx + c$	α and β		

Cubic Function	Roots	Sum of Roots	Product of Roots	Product Pair and Sum
$y = ax^3 + bx^2 + cx + d$	α, β, and γ			

What is the pattern?

Polynomial Function	Roots	Sum of Roots	Product of Roots

In pairs, write down in words the relationship between the sum of the roots and the coefficients for any polynomial. (Generalizations)

FIGURE 5.3: AN EXAMPLE OF A GUIDED INQUIRY TASK

Sum of the Roots for Any Polynomial
Product of the Roots for Any Polynomial

Investigate the relationship between the sum of the roots and the coefficients of any polynomial.

Investigate the relationship between the product of the roots and the coefficients of any polynomial.

Explain any generalizations you discover and include examples to illustrate how your generalization works.

© 2014 Jennifer Wathall

your class listening passively for extended periods of time, provide opportunities for your students to engage from the very beginning of a lesson. David Sousa (2015) recommends that lessons should be taught in 15- to 20-minute segments, as they are optimal times for working memory to be able to focus and retain learning in adolescences. This suggests that for optimal learning, 60-minute lessons should be broken into three parts.

Thomas Warren-Price (2003) concluded that less TTT resulted in more student autonomy. Reducing TTT will encourage student-centered learning and engagement and reduce the risk of students' inattention. When you are talking at length at the front of the room, how do you know your students are listening or even learning? My aim in every lesson is to whole class talk no more than 20%–30% of the time. This does not mean you are not talking to individuals and groups of students for the majority of the lesson. Hattie (2009), in his meta-studies, has found the effectiveness of TTT satisfactory for introducing new information but a poor route to deeper learning.

Strategy 5: Cater to Everyone in Your Class; use Differentiation Strategies

When your student answers questions correctly, how do you know whether she has learned something new or whether she knew that before? How do you know

FIGURE 5.4: STUDENT'S RESPONSE TO A STRUCTURED INQUIRY TASK

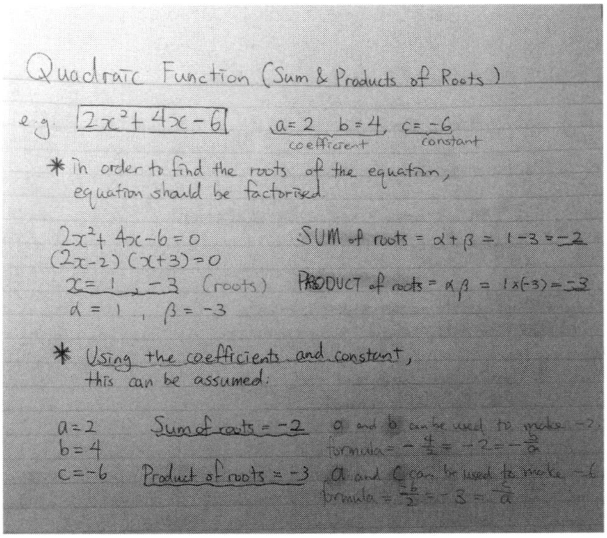

Quadratic Function (Sum & Products of Roots)

e.g. $\boxed{2x^2 + 4x - 6}$ $a = 2$ $b = 4$, $c = -6$
                              coefficient   constant

* In order to find the roots of the equation,
   equation should be factorised.

$2x^2 + 4x - 6 = 0$ SUM of roots = $\alpha + \beta = 1 - 3 = -2$
$(2x - 2)(x + 3) = 0$
$x = 1, -3$ (roots)   PRODUCT of roots = $\alpha\beta = 1 \times (-3) = -3$
$\alpha = 1, \beta = -3$

* Using the coefficients and constant,
   this can be assumed:

$a = 2$ Sum of roots = -2 a and b can be used to make -2
$b = 4$ formula = $-\frac{4}{2} = -2 = -\frac{b}{a}$
$c = -6$ Product of roots = -3 a and c can be used to make -6
                                     formula = $-\frac{6}{2} = -3 = \frac{c}{a}$

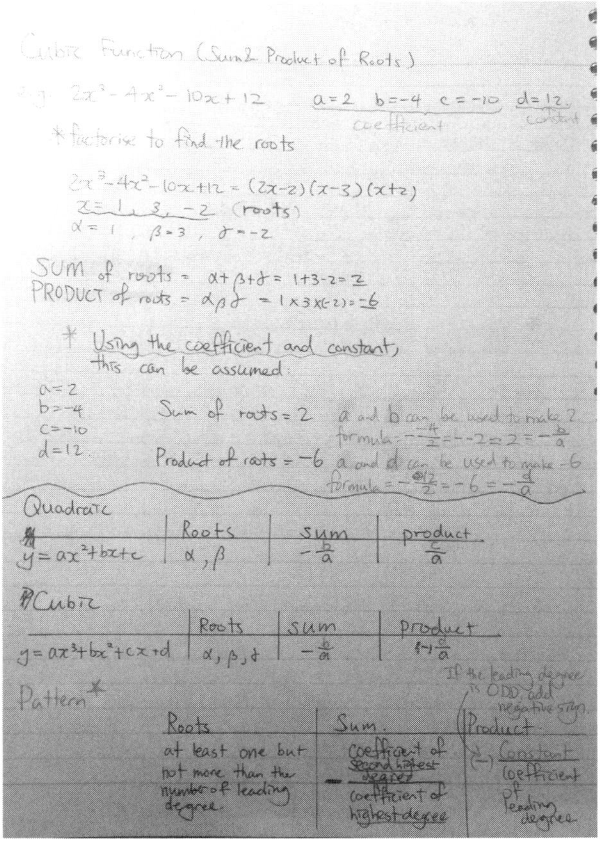

Cubic Function (Sum & Product of Roots)

e.g. $2x^3 - 4x^2 - 10x + 12$ $a = 2$ $b = -4$ $c = -10$ $d = 12$
                                       coefficient   constant

* factorise to find the roots

$2x^3 - 4x^2 - 10x + 12 = (2x - 2)(x - 3)(x + 2)$
$x = 1, 3, -2$ (roots)
$\alpha = 1, \beta = 3, \gamma = -2$

SUM of roots = $\alpha + \beta + \gamma = 1 + 3 - 2 = 2$
PRODUCT of roots = $\alpha\beta\gamma = 1 \times 3 \times (-2) = -6$

* Using the coefficient and constant,
   this can be assumed:

$a = 2$
$b = -4$ Sum of roots = 2 a and b can be used to make 2
$c = -10$ formula = $-\frac{-4}{2} = -2 = 2 = -\frac{b}{a}$
$d = 12$ Product of roots = -6 a and d can be used to make -6
                                      formula = $-\frac{12}{2} = -6 = -\frac{d}{a}$

Quadratic

$y = ax^2 + bx + c$	Roots	sum	product
	α, β	$-\frac{b}{a}$	$\frac{c}{a}$

Cubic

$y = ax^3 + bx^2 + cx + d$	Roots	sum	product
	α, β, γ	$-\frac{b}{a}$	$-\frac{d}{a}$

If the leading degree
is ODD, add
negative sign.

Pattern *

Roots	Sum	Product
at least one but not more than the number of leading degree	$\dfrac{\text{coefficient of second highest degree}}{\text{coefficient of highest degree}}$	$\dfrac{\text{constant}}{\text{coefficient of leading degree}}$

every student in your classroom is learning and making progress? Effective **differentiation** strategies ensure all students make progress in their learning, regardless of their starting point. Carol A. Tomlinson suggests an alternative definition for differentiation is "respectful teaching." A common strategy for differentiation in the math classroom is to give more able students more work and less able students less work. Is this an equitable way to differentiate students? Most likely not. Instead, teachers can offer students various choices to approach the curriculum, all of which are equally challenging and equally time-consuming. Visit the companion website for some examples of differentiation and assessment strategies (Figures M6.1–M6.4).

Tomlinson and Imbeau (2010) outline four key curriculum-related elements of a differentiated classroom, which are based on three categories of student need and variance (readiness, interest, and learning profile):

1. Content: Giving tiered tasks of varying levels of difficulty and allowing students to choose entry levels represents differentiating content, in mathematics.

Differentiating content implies that students do have varying entry points and allows every student to experience progress and challenge when learning math. This is probably the most common way of differentiating in a math classroom.

2. Process: Differentiating the process implies that the learning outcomes are the same but students are given a choice of process during the learning experience.

An example of differentiating the process could be setting up expert groups in a classroom, with students determining which stations to visit based on need. Different expert groups at each station could have a different approach when presenting the ideas and concepts to other students. Some stations could use videos on various methods of solving polynomials, while other stations may ask students to engage in a game or activity.

Structured and guided inquiry tasks are also methods to differentiate the process while maintaining the same learning outcomes. The highly scaffolded structured inquiry approaches can cater to students who need more prompts and guidance while guided inquiry approaches feature fewer prompts.

Another example is Sternberg's and Grigorenko (2000) **tri-mind strategy**, in which students are given the same learning outcomes but three different ways to approach them: analytically, practically, or creatively. This is a tool to support differentiation of different thinking styles in your classroom. Tri-mind activities will be discussed in detail in the next section under practical strategies. Figure 5.5 illustrates a tri-mind activity.

3. Product: The product is what a student essentially produces at the end of a unit to demonstrate mastery of the content.

Give students a choice for the medium of their product: video, website, presentation from a tiered activity. Authentic performance assessments, in Chapter 6, are also examples of giving students the opportunities for a differentiated product.

4. Affect: Affect refers to the ability to move emotionally or to touch the feelings of your students.

If I think about the most revered and respected teachers in my school, second to their content knowledge is their ability to build strong, positive relationships with their students. Emotions affect the ability to learn, motivation for a subject, and one's self-esteem. An important lesson I have learned over the years is you don't know what your student has experienced at home the night before or what personal challenges he or she may be facing in life in general. The classroom should be a safe, secure, and positive environment where students are able to build their confidence and feel unconditionally accepted. Overwhelming research supports that knowing your students and connecting on a positive level facilitates increased learning and motivation. I try to form bonds and connections with each and every one of my students so they know I care about their progress and care about them as a human being. Computers will never be able to replace teachers for this reason.

Practical Strategies to Differentiate Your Classroom

Tri-Mind Activities

Sternberg's and Grigorenko (2000) work on the tri-mind presents three different thinking styles: analytical, practical, and creative: "Many students could learn more effectively than they do now if they were taught in a way that better matched their patterns of abilities. Teaching for successful intelligence provides a way to create such a match. It involves helping all students capitalize on their strengths and compensate for or correct their weaknesses. It does so by teaching in a way that balances learning for memory, analytical, creative, and practical thinking" (p. 274).

Analytical thinkers enjoy facts and figures, breaking down things, and understanding how things work.

Practical thinkers appreciate context and real-life settings. They are the doers who like to organize and get things done. They prefer hands-on activities.

Creative thinkers like to be empowered to produce, create, and design. They often have a unique perspective and sometimes enjoy working alone.

Students have a preferred thinking style—analytical, practical, or creative—when learning, but it is equally important to give students opportunities to employ different thinking styles and to encourage students to work in a thinking style that is not their preferred.

Figure 5.5 is an example of Sternberg's tri-mind activity in which students are given choices about how to approach a summary on functions at the end of a unit. For

FIGURE 5.5: TRI-MIND ACTIVITY ON FUNCTIONS

End of Unit Evaluation on Functions

1. Choose one of the options below.
2. Find two other people who have also made the same choice as you.

Choice 1	Choice 2	Choice 3
Make a summary and explain all the key concepts in the topic of functions.	Explain how you would use all the concepts in functions to answer questions. Include lots of examples of "how" you answer questions.	Create a graphic (visual) or metaphor of how all the concepts of functions are related. Provide examples to illustrate your points.
My choice is:	I chose this because:	
The members of my group are:	Outline of individual roles and action plan	

3. Work on your task together and be ready to share.

additional resources on tri-mind activities, see Figures M6.6, M6.7, and M6.8 on the companion website.

Figure 5.6 is an example of an analytical approach from my Year 10 (Grade 9) class.

FIGURE 5.6: STUDENTS' EXAMPLE OF AN ANALYTICAL APPROACH TO FUNCTIONS

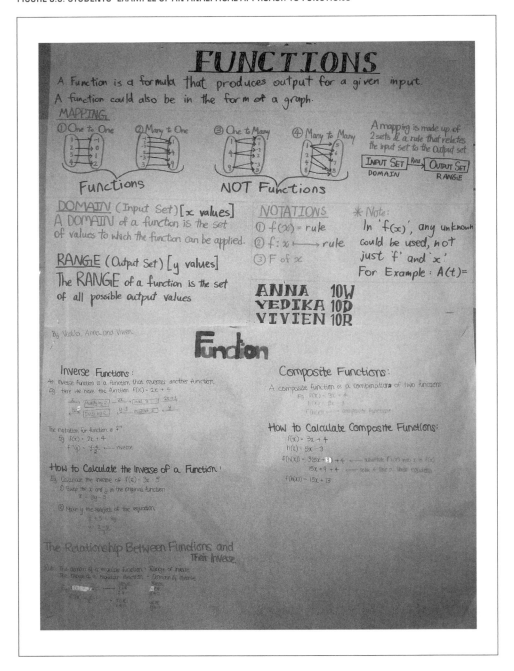

Anna, Vedika, and Vivienne explain the unit on functions clearly and logically. This response illustrates a thorough and deep understanding of all of the components that make up this unit.

Hint Jar or Hint Cards

Another strategy to encourage differentiation in the classroom is the use of "hint cards" or a "hint jar." Students can choose to take these hint cards when they need some guidance to move on to the next stage or to understand a concept better at any time during a lesson.

Figure 5.7 shows examples of hint cards for the topic of functions. One side of the card has the concept and the other side has conceptual explanations.

Giving Students Choice

Nunley (2004, 2006, 2011) summarizes the three steps to layering curriculum to facilitate differentiation:

1. Add some choice
2. Hold students accountable
3. Encourage higher-level thinking

The first step in Nunley's model focuses on giving students choice in lessons. This automatically motivates students by giving them a sense of control. Nunley suggests the following strategies for adding some choice:

> **Give students choice to motivate and engage.**

- Allow students to practice the understanding of concepts through different mediums such as practicing problems through computer games, using manipulatives to complete a task, or working in small groups
- Give students choice whether they would like to attend a mini workshop with you to clarify conceptual understandings
- Allow students to choose from a variety of different tasks offered
- Allow students to choose whether they need to watch a video to support learning

Another strategy to give students choice and to provide differentiation is to use Hyde's (2006) K–W–L, which focus on three questions (see Figure 5.8).

The visible thinking routine "Question Sorts" (Harvard University, Project Zero, 2007) also allows students choice in their learning by asking students in groups

FIGURE 5.7: HINT CARDS FOR FUNCTIONS

Functions

Functions represent mappings which are

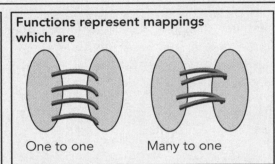

One to one   Many to one

Functions

Functions represent mappings which are
One to one   Many to one

Here are some analogies that show the relationship of functions:

Personal login to a website and password
How many days in a month and the actual month
Heights and each person's unique fingerprint
Hair color and signature

Functions

Functions represent mappings which are
One to one   Many to one

The vertical line test can help us to determine whether a rule is a function or not.
For a function, no vertical line should cross more than once with a graph.

Functions

Vertical Line Test

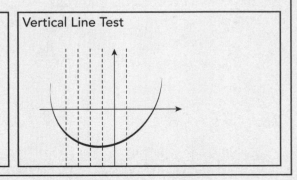

K	What do I know?
W	What do I want to learn more about?
L	What did I learn?

to brainstorm a list of questions about a particular unit and prioritize and address each question. This allows students to identify questions to direct future inquiry and deepen understanding.

The second step in Nunley's model—hold students accountable—is concerned with awarding grade points for the actual learning of an objective or the acquisition of a particular skill rather than completion of an assignment or for practicing a skill. Many courses run on the basis that students can pass courses by completing assignments without having learned anything at all. Focusing on actual learning will support individual student progress and promote differentiation.

The third step in Nunley's "layered curriculum" involves encouraging more complex learning by dividing the instructional unit into three layers:

a. Basic rote information (layer C)

b. Application and manipulation of that information (layer B)

c. Critical analysis of a real-world issue (layer A)

These layers align with the philosophy behind the levels in the Structure of Knowledge and the Structure of Process: the concept-based, three-dimensional model for curriculum and instruction.

Strategy 6: Assessment Strategies

How do you know your students are learning? I have posed this question to many teachers and often receive this answer: "If I ask them about the work and they can tell me about it."

How do we know they didn't know this before? Diagnostic (pre-assessment) and formative assessment strategies can be used before, during, and at the end of a lesson or at the end of a topic or unit. These types of assessments provide feedback to teachers for future instructional planning and identify what your students already know.

One of my favorite strategies that gives me immediate feedback during a lesson and encourages learners to think on a metacognitive level is the "learning curve" card illustrated in Figure 5.9. This tool has been adapted from Michael Fullan's (2008) "implementation dip" theory.

Here are two ways to use the learning curve tool:

1. Students are asked to create a symbol representing themselves (for anonymity), and they are asked to attach their symbol, using a reusable adhesive (putty or blu-tack), to a giant poster version of the dip curve (Figure 5.9) displayed in the classroom. Students are encouraged to move their symbol at any stage of their learning, during a lesson or over a series of lessons. If students start at the right side (end) of the dip, indicating they already understand the concepts of a topic, then instruction needs to be modified and differentiated to ensure all students are captivated by the lesson and making progress in their learning.

2. Each student has his or her own mini laminated version of the dip curve and is given a reusable adhesive. They place their reusable adhesive anywhere on the curve at the beginning of the lesson and move at any time during the lesson. When I walk around, I can very efficiently see who is struggling and who needs help.

Several other pre-assessment, formative, and summative assessment strategies, such as exit and entrance tickets, are outlined in more detail in the next chapter.

FIGURE 5.9: THE LEARNING CURVE: WHEN YOU LEARN SOMETHING NEW . . .

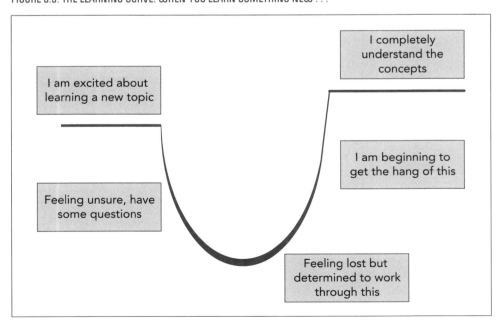

Strategy 7: Be Purposeful When Asking Students to Answer Questions; There Is Safety in Numbers

I have never been a believer in "hands up" when asking for feedback from a group discussion. The same students answer, and the quieter students sit back and let others do the work. When you randomly call upon students and employ techniques such as the paddle pop stick technique, you run the risk of your student not knowing the answer and feeling publicly embarrassed. In the paddle pop stick technique, each student's name is assigned a paddle stick, and teachers draw students' names randomly from paddle pop sticks during the lesson. I have observed lessons where students sit in class terrified, waiting for their name or number to be called out on their paddle pop stick. This can be a very demotivating strategy, especially for students who have very little confidence in their mathematical ability, and it can lead to increased math anxiety.

> Often math anxiety and fear is encouraged when students are called out randomly to answer questions.

One technique I do employ is "thumbs at chest." When students have completed a task or know the answer to a question, they put their thumb to their chest, which provides a more discreet way for teachers to see who has understood and who has not. If I see students without their thumb to their chest, or even with their thumb down, this gives me a discreet way of knowing who needs more support during a lesson.

In my opinion, the most effective way to get feedback during group discussions is to ask for paired or group feedback. Students feel safety in numbers and can rely on their partner or group members for support and learn simultaneously if they do not understand a concept. This ensures that every student's opinion is shared with someone in the class in a less daunting way.

One of my favorite partner talk activities is from Harvard University's Project Zero (2007), "Think, Pair and Share." This involves posing a question and asking students to think independently first. After a few minutes, ask students to pair up and discuss their ideas. Last, ask students to share their feedback with the group. Not only do you make thinking visible in your classroom, you also give students the opportunities to discuss and share their ideas, creating a social learning environment.

Strategy 8: Flexible Fronts: Arranging Your Classroom

Traditional classrooms are arranged with a front of the class and a back. The students who try not to be noticed tend to sit at the back, with eager beavers sitting at the front of the room. Do we really need to have a front and a back of a classroom? Structuring your learning environment is one of the key factors for effective collaboration.

Modern college classrooms are configured completely differently from classrooms of the 1950s. In the article "Innovation Enters the Classroom" (n.d.), Ron Briggs, senior lecturer and coordinator of general chemistry at Arizona State University (ASU), has restructured his learning spaces to move away from the "sage on the stage" to learning spaces that promote group work and facilitation. Learning spaces include projection on multiple screens and furniture that supports collaboration. This arrangement supports the social constructivist teaching and learning approach. This learning space design puts the teacher anywhere in the space and immerses the teacher in the learning environment. There is no particular front of the room. As a result of this classroom set up, students are more engaged and coach each other. This change in structure has resulted in an increase of retention levels at ASU by nearly 5%, and grade performance has increased 3% to 4% at the university. ASU is one of the largest universities in the U.S., and budgets and resources are tight. This new configuration costs less to operate than the traditional classroom due to the more efficient use of the space and more effective learning. There is also a reduced need for teaching assistants.

The Queensland University of Technology (QUT, 2014) in Australia provides learning spaces that promote collaborative learning and practical problem solving. The QUT website states:

> "Learning spaces are equipped with:
>
> - furniture that can be configured to suit different classes and activities
>
> - built-in technology that facilitates interaction between students and lecturers, like lecterns with screen-sharing capabilities and computers on wheels (COWs) for student group work.
>
> This makes it easier for students to participate in discussions, share ideas, and ask and answer questions."

The physical set up of these learning spaces is similar to that in Figure 5.10. I have trialed a flexible front classroom with the configuration shown below in Figure 5.10. I noticed this arrangement is more conducive for group work, encourages higher levels of engagement from the students, and I float from group to group, having conversations with one or a group of students throughout the lesson.

And that is how change happens. One gesture. One person. One moment at a time

—Libba Bray, 2007.

Providing an environment conducive for learning to ensure conceptual understanding starts with one tool, one strategy, one step at a time.

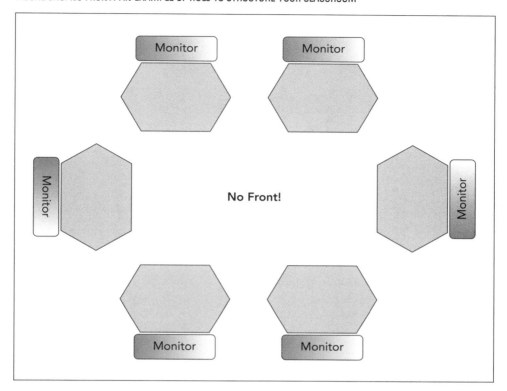

Typical Lesson Using Flexible Fronts

Instructions to students:

We have completed a unit of study on trigonometry. In your groups, make a summary of the key points of the unit and also create a real-life problem that requires trigonometry to solve it. Ensure that each group member has a specific role and responsibility. Use either a Google Presentation or Google Doc to allow everyone to contribute to the summary.

Teacher's notes:

Students sit in groups of six with one monitor and share one Google Doc or Google Presentation. One student connects his or her computer to the monitor so everyone in the group can see the work in progress (and this allows me to check progress also). Once groups complete the learning experience above, they present to the rest of the class from their monitor.

Chapter Summary

This chapter outlines eight strategies for an engaging, inquiry-led, concept-based classroom:

1. Encourage learning through group work and allow students to discuss at every opportunity through collaboration.

2. Provide an open, secure environment to allow mistakes to be used as the learning process by fostering a growth mindset.

3. Use appropriate levels of inquiry (structured, guided, or open) according to your students' needs and employ inductive approaches to develop conceptual understanding. The exemplar about polynomials included in this chapter illustrates the inductive teaching approach—using the levels of inquiry and emphasizing synergistic thinking between the factual and conceptual levels of knowledge and understanding to draw the understandings from the students, rather than telling them what they should understand.

4. Reduce whole class talk time from the teacher (TTT) and plan to engage your students in 15–20 minute blocks.

5. Ensure your lesson includes differentiation strategies by looking at content, process, product, and affect to cater to all of your students' needs.

6. Collect diagnostic and formative feedback regularly before, during, and after the lesson.

7. Be purposeful when calling upon a student; no random choice of students. Instead, ask for group or paired feedback.

8. Flexible fronts! Structure your classroom so the teacher is integrated in the learning and acting more as a "guide on the side" and a "meddler in the middle" rather than a "sage on the stage."

The next chapter will look at assessment strategies to appraise students' conceptual understanding.

Discussion Questions

1. What strategies do you employ to engage and motivate the hearts and minds of your students?

2. How do you keep students on task in a social learning environment?

3. How do you promote a growth mindset in your classroom and encourage mistakes as part of the learning process?

4. Do you think it is important to balance whole teacher talk time with student working time? Why or why not?

5. What differentiation strategies do you use to cater to the individual needs of your students?

6. What formative assessment strategies do you employ in your classroom?

7. What role do you think the structure of your classroom plays in student learning?

....

CHAPTER 6

How Do I Know My Students Understand the Concepts?

Assessment Strategies

As a workshop leader, I facilitate continuing professional development and I love to share this quote:

> *Everybody is a genius. But if you judge a fish by its ability to climb a tree, it will live its whole life believing that it is stupid. (Unknown)*

It is vital that every student feels math is accessible and that, with effort, he or she can increase his or her understanding and achievement in math. Different modes of assessment can help individual students feel valued and help them gain confidence in their ability. Assessment design needs to cater for all abilities as well as differentiate for individual student needs. Assessment comes from the Latin *assessus*, past participle of *assidēre*, which means "to sit beside." Keeping that in mind, the purpose of assessment should be to inform teaching and learning and "sit beside" the student by guiding him or her. Assessment can be used as a diagnostic, formative, or summative tool.

Wiggins and McTighe (2006a) talk about the twin sins in traditional curriculum. One sin is an activity-oriented design in which activities do not lead to any intellectual challenge or development. They call these activities "hands on without being minds on." These types of time-filling activities do not focus on authentic conceptual understanding. The second sin is one of coverage of content with little understanding of the concepts by the learner. The description of education as "the kindling of a flame, not the filling of a vessel" is often attributed to Socrates, though the quote may well have come from Plutarch. Why is it, then, in this day and age, that coverage is still equated to learning and understanding when the problem of coverage was recognized more than 2000 years ago?

Students need to be exposed to a variety of different forms of assessment to improve understanding and achievement. Traditional mathematics assessment includes summative, computational, closed questions, which tend to assess memorization of facts and processes rather than conceptual understanding. More examination boards worldwide are trying to move toward questions that assess for conceptual understanding, which require application of understanding and encourage collaboration through group assessments.

Examinations are an integral part of education, but we do not have to "teach to the exam." The journey to that end point can be invigorating and intellectually challenging while developing creativity, collaboration, and communication skills. Multiple means of assessing give students the opportunities to develop key skills and promote conceptual understanding of the content before, during, and after a unit of work. Assessment also informs future curriculum planning, and the ultimate goal is to improve student learning and achievement.

Before starting any unit of work, a vital part in differentiating instruction is the use of pre- or diagnostic assessment. Identifying different students' pre-existing knowledge helps inform instruction. One of my favorite activities is "Write down everything you know about . . ." This identifies what students do know quite effectively and steers the focus away from what they don't know. Other advice for pre-assessment includes the following:

- Keep pre-assessment brief
- Emphasize HLY (haven't learned yet) rather than what students do not know
- Design pre-assessments that involve active learning (Laud, 2011)

The next few sections will discuss a variety of assessment strategies with the goal of informing instruction.

Assessments With Conceptual Depth

Inquiry-Based Assessment

Ollerton and Vasile (2014) discuss inquiry-based assessments, referring to them as "fluffy" assessment strategies. We usually think of the term "fluffy" to mean lacking depth and rigor. Ollerton and Vasile, however, use this term to mean nontraditional, inquiry-focused assessment. A colleague who was against these types of inquiry-based assessments coined the term "fluffy" for these assessments initially but soon realized their value for checking understanding.

Inquiry-based assessment evaluates conceptual understanding and is often categorized as specific conceptual questions. These types of assessments include open inquiry-style questions where there are a number of possible outcomes that may be unpredictable. Mike Ollerton and Daniela Vasile (2014) write, "Importantly an EBL (Enquiry Based Learning) form of assessment enables the teacher to see the kind of conceptual connections students form between different areas of mathematics, i.e. angle, tessellation, and factors. This is an important part of learning mathematics."

Figure 6.1 shows examples of inquiry-based questions.

These examples of inquiry-based assessment questions and prompts ask students to describe the conceptual understandings involved to reflect understanding and purpose of what they are learning.

Inquiry-based assessments can be used in three ways:

- At the beginning of a unit as a diagnostic tool, to seek prior knowledge;
- During a unit of work as formative assessment, to inform about student understanding, progress, and the focus of instruction for improvement;
- At the end as a summative tool, to assess student understanding and achievement.

Another benefit of inquiry-based assessments is the development of mathematical communication skills.

Open Inquiry Tasks and Open-Ended Questions

The International Baccalaureate (IB) mathematics standard and higher level courses currently include an internal assessment component called the "personal exploration" (International Baccalaureate, 2014). The personal exploration is an example of an authentic open inquiry task in which every student is encouraged to take his or her own line of inquiry, with completely unpredictable outcomes. Students choose their own topic of interest and line of inquiry. The personal exploration is a criterion-referenced type of assessment with

1. Write down everything you know about right-angled triangles.

   This type of prompt is excellent for finding out previous knowledge, misconceptions during a unit of work, and what students have learned at the end of a module.

2. Explain sin x, cos x, and tan x. Include in your explanation the conceptual relationships.

   Most students give the reply, "They are buttons on a calculator." Authentic conceptual understanding of these trigonometric ratios comes from understanding the concepts of similarity, right angle, triangle, and ratios.

3. How would you work out sin x = ½ without a calculator?

   Similar to the goal of question 2, students can give graphical or analytical explanations here to show conceptual understanding.

4. Explain the Pythagorean theorem and include an example to illustrate your explanation.

   Rather than memorizing a fact ($a^2 + b^2 = c^2$), students need to understand the conceptual relationships of area, triangle, right angle, and squaring.

5. When is the cosine rule used? Give an example. When is the sine rule used? Give an example.

   Students need to know how to apply this rule and also determine in which situations it is more efficient to use the cosine rule. This requires critical thinking skills and the ability to problem solve effectively.

6. Here are three similar right-angled triangles. Write down everything you know about them.

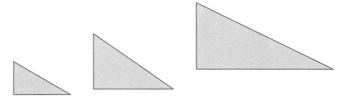

   For this prompt, have students talk about the Pythagorean theorem or sine, cosine, and tangent of angles.

7. Explain what a radian is.

   Rather than focus on the process of converting an angle from degrees to radians, students need to be able to understand the concepts of arc length, angle subtended, and radius for further conceptual engagement.

five criteria: communication, mathematical presentation, personal engagement, reflection, and use of mathematics. For authentic learning, students need to be able to communicate their mathematical ideas in a coherent fashion. The goal of the personal exploration is to provide students with opportunities to pursue areas of mathematics that are not covered by the examination and to cultivate an appreciation of mathematics beyond the final examination. My students have explored and enjoyed all sorts of interesting topics, such as looking at the math involved in architecture, basketball, and even which modes of transport are safer. Future curriculum and instruction plans need to include more opportunities for this type of student-directed assessment to encourage independent, enthusiastic lifelong learners.

Conceptual understanding is not just an accumulation of knowledge; it is an ability to transfer and apply knowledge to new, unfamiliar situations. Opportunities to answer open-ended questions in mathematics require complex thinking on a deeper conceptual level. Open-ended questions also allow for differentiation of abilities. Here are some examples of open-ended questions with multiple, unpredictable outcomes:

- One value for θ is $\frac{\pi}{6}$. Name five other possible values if $sin\,\theta = 0.5$.
- Use any of the information on the cards (you can repeat cards, you have to use at least one of the cards, and you may write anything on the blank card) to write a correct and incorrect equation.

tan	θ	4	3	π	

- Choose values for the parameters a, b, c, and d and sketch with an explanation of the effect of each parameter: $y = a\,\sin(b(x - c)) + d$.

- Investigate one of the many Pythagorean proofs and explain.

Visible Thinking Routines

Many classrooms utilize visible thinking (Harvard University, Project Zero, 2007), which can be useful for diagnostic and formative feedback. It allows students to express and communicate their personal thoughts and ideas, giving teachers an insight to student's thinking. Figure 6.2 shows some adapted examples of visible thinking strategies.

The "three do's and three don'ts" can be used for different topics, such as three do's and three don'ts when solving an equation.

FIGURE 6.2: EXAMPLES OF VISIBLE THINKING ROUTINES

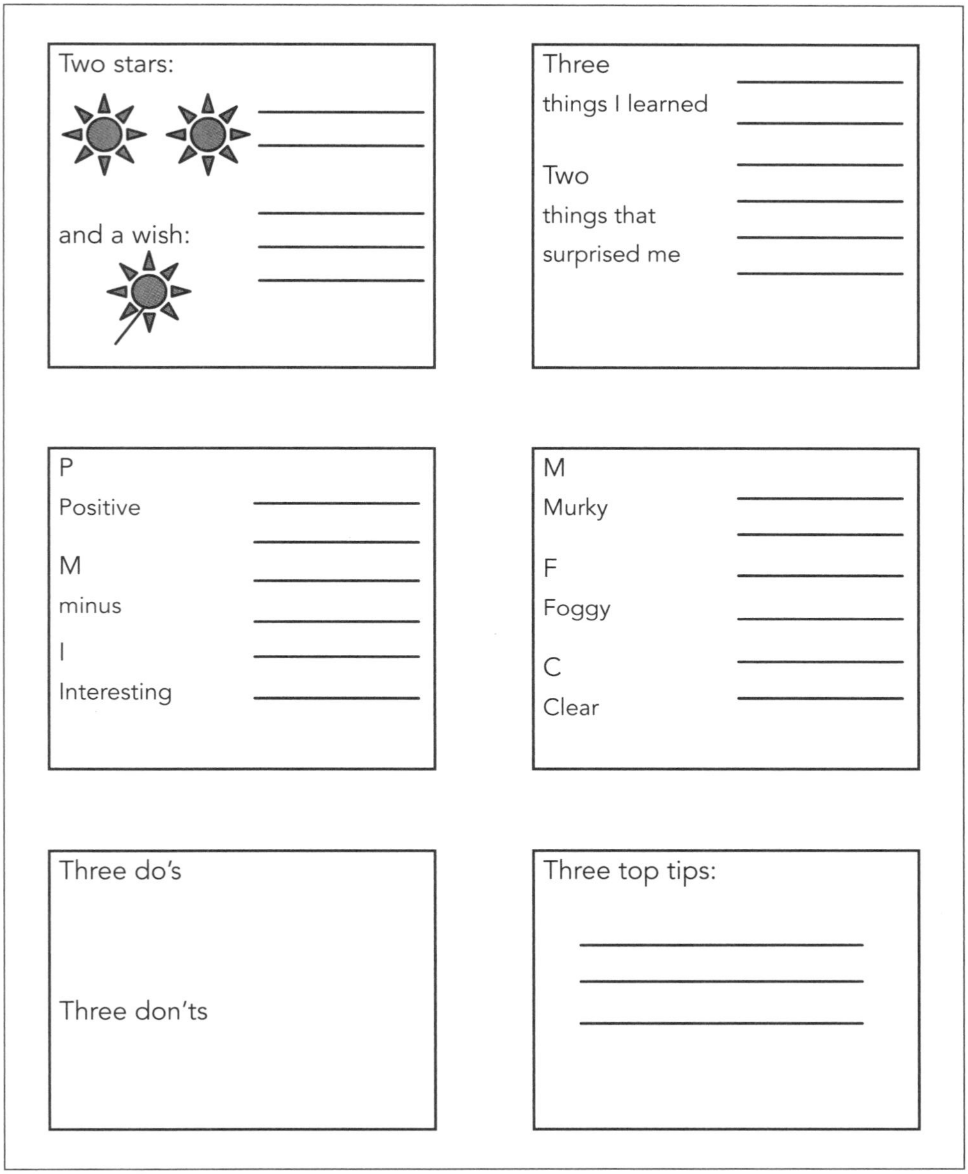

Two stars:

━━━━━━
━━━━━━

and a wish:

━━━━━━
━━━━━━
━━━━━━

Three things I learned

━━━━━━
━━━━━━

Two things that surprised me

━━━━━━
━━━━━━
━━━━━━

P
Positive ━━━━━━

M
minus ━━━━━━
━━━━━━

I
Interesting ━━━━━━

M
Murky ━━━━━━
━━━━━━

F
Foggy ━━━━━━

C
Clear ━━━━━━

Three do's

Three don'ts

Three top tips:

━━━━━━
━━━━━━
━━━━━━

Do keep equations balanced by doing the same operations to both sides of the equation.

Don't forget to think about PEMDAS in reverse when solving an equation.

The three don'ts can also be used to identify classic mistakes that students make.

These routines may be used as entrance tickets at the beginning of a lesson, which can be used to gauge previous knowledge and prior learning to help determine the focus for the lesson. Often entrance and exit tickets may be interchangeable and may also be used during a lesson.

Figure 6.3 shows more ideas for visible thinking routines that may be used interchangeably at the beginning or end of a lesson.

FIGURE 6.3: MORE EXAMPLES OF VISIBLE THINKING ROUTINES

Write down what you think _____ is. Describe in your own words the meaning of _____.	What would you like to learn about _____? Why do you think this is important?
Write down three misconceptions about _____.	Write down three things you know about _____. Write down three things you would like to know about _____.

Project Zero (Harvard University, 2007) and the work of Ritchhart Church, and Morrison (2011) outline several excellent visible thinking routines. The following routines encourage synthesizing and organizing ideas.

- Think, Pair and Share: This allows students time to think, discuss with a partner, and then share to the rest of the group. This allows you to assess student thinking around a particular concept.

- Connect, Extend, Challenge: This is a routine that allows students to connect their prior knowledge and understandings to new ideas. Figure 6.4 outlines questions posed in connect, extend, challenge.

- Headlines: I love using this at the end of a lesson or unit. I ask students to write down a headline (similar to a newspaper headline) that

Connect:	How are these ideas CONNECTED to what you already knew?
Extend:	What new ideas did you get that EXTENDED or pushed your thinking in new directions?
Challenge:	What is still CHALLENGING or confusing for you to get your mind around? What questions, wonderings, or puzzles do you now have?

Church et al. (2011, p. 132).

summarizes the key ideas and concepts of the lesson or unit. The headline lets me know immediately if students have not understood the essence of the learning experience and what I need to revisit and with whom. Here are some examples of headlines about trigonometry:

SOHCAHTOA helps to solve real-life problems by finding unknown lengths and angles in right-angled triangles.

A new discovery! All similar right-angled triangles share the same ratios.

- Generate, Sort, Connect, and Elaborate: This helps students to connect concepts through a concept mapping exercise. An example of this would be to ask students to generate all the key concepts in trigonometry. They may come up with a variety of concepts that could be sorted into non-right-angled and right-angled trigonometry, and this informs me whether students understand the concepts involved in right-angled and non-right-angled trigonometry. Within right-angled trigonometry, students then have to find connections between the key ideas and elaborate on them.

- I used to think… Now I think…: This routine helps me to see what students' misconceptions were before learning or how they changed their thinking over the course of the unit. The goal is to see how students' thinking has deepened, shifted, or changed. Here are some examples of this routine:

I used to think that I could use SOHCAHTOA for any triangle.

Now I think it can only be used for right-angled triangles.

David Sousa (2015) describes cognitive closure as the process when students' working memory summarizes for itself the perceptions of what has been learned. Several visible thinking routines can be used for cognitive closure. This is a vital part of a lesson when a student attaches sense and meaning to new learning and thereby increases the chances learning will be retained in long-term memory. Cognitive closure can be conducted at various times in a lesson, but it should almost always be conducted at the end of a lesson. Cognitive closure is set apart from a review of a lesson by the teacher as closure involves the students thinking and making sense of the concepts and deciding whether they make sense. Closure also gives students opportunities to ask questions to clarify any misunderstandings. In the words of David Sousa, "Cognitive closure is a small investment in time that can pay off dramatically in increased retention of learning" (p. 199).

Performance Assessment Tasks

All literature around Concept-Based Curriculum and Understanding by Design (UbD) clearly states the difference between "time filling" activities with no intellectual engagement and designing authentic performance tasks to assess student understanding.

> **Information is most likely to get stored if it makes sense and has meaning.**
> **(Sousa, 2015)**

Performance assessment tasks assess both process and product and involve students constructing various products with an audience in mind. Quality performance assessment tasks engage student intellect and develop conceptual understanding in a real-world context. They can be a single lesson or a series of lessons. Sousa (2015), in his brain research, states that information is most likely retained if it makes sense and has meaning.

Performance tasks assess what students understand (generalizations), what they know (factual knowledge), and what they can do (skills) in a particular unit.

Wiggins and McTighe (2006a) define performance tasks as "complex challenges that mirror the issues and problems faced by adults" (p. 153). They also list characteristics of authentic tasks:

- Realistically contextualized

- Requires judgment and innovation

- Asks student to carry out work and exploration in a discipline that resembles work done by people in that field

Erickson and Lanning (2014, p. 88) list the characteristics of a quality performance task:

- Aligned with know, understand, and able to do

- Authentic or scenario based

- Evaluated against clear criteria

- May offer student choice

To provide a framework when designing authentic performance tasks, a commonly used tool is RAFTS (Holston & Santa, 1985). This is a tool to design projects and assignments that was originally formulated to help students understand language and literature from novels but can be applied across disciplines to any performance task design. RAFTS is an acronym:

Role: the role the student will take during the project. In the performance task above, the role of the student is designer.

Audience: the targeted people who will read the project. In the example below in Figure 6.7, the audience is the architects who will be rebuilding Island School.

Format: This could include different modes such as newspaper article, design proposal, brochure, or documentary.

Topic: the subject matter. This usually encompasses the conceptual understandings of the topic.

Strong: verbs and adjectives to provoke interest and excitement.

Wiggins and McTighe (2006a) describe a similar framework designing tool called "GRASPS." GRASPS stands for Goal, Role, Audience, Situation, Performance, and Standards.

The structure of the task planner—What, Why, How—promotes the development of quality performance tasks (Erickson, 2007). This complements the RAFTS model to help students understand the What, Why, and How concerning the task. Figure 6.5 illustrates a math performance task for the topic "Linear Functions in Real Life." The questions—What, Why, How?—support the development of the task. Figure 6.6 shows an example of the RAFTS model and how this framework is used for a math performance task.

Figures 6.7 through 6.9 are examples of one mathematics performance task, consisting of a series of lessons on trigonometry.

Figure 6.7 sets the scene and presents a real-life scenario. The task is to measure all the heights of all the buildings in the school using trigonometry and to represent the current school in a scale drawing. The final part of the task is to design a new school using the same footprint.

FIGURE 6.5: MATHEMATICS PERFORMANCE TASK ON LINEAR FUNCTIONS

What	Investigate (unit title or key topic)
	Linear functions in real life
Why	In order to understand that (generalization)
	Variables connected in direct proportionality imply a linear relationship.
	A function represents a mapping (rule) that assigns each input (domain) with one output (range).
	Linear functions show relationships that exhibit a constant rate of change.
	Transformation (translation) of a linear function changes the rate of change and the initial value.
How	Performance task/engaging scenario for students:
	You are a reporter who is investigating how real-life situations can be modeled by linear functions and the importance of how they are used for different purposes. Create a documentary explaining the uses of linear equations and how they are used to solve real-life problems.

Adapted from Erickson (2002).

FIGURE 6.6: USING THE RAFTS MODEL TO DESIGN A PERFORMANCE TASK

Unit Title/Key Topic: Linear Functions in Real Life

R = Role	Investigative reporter
A = Audience	Peers
F = Format	Video in the form of a documentary
T = Topic	Linear functions
S = Strong verbs and adjectives	Create, solve

Adapted from Holston and Santa (1985).

Figure 6.8 is an example of an inductive inquiry activity, which helps students to understand the trigonometric ratios of similar right-angled triangles.

Figure 6.9 shows students how to use trigonometry to measure the heights of buildings by going through the calculations for one of the buildings.

For more examples of performance tasks (Figures M7.1 and M7.4) and templates for designing performance tasks (Figures M7.2 and M7.3), visit the companion website. You will also find a list of task formats and roles there.

FIGURE 6.7: PERFORMANCE TASK: SETTING A SCENARIO

The Rebuilding of Island School, Hong Kong

Island School was built in its current location in 1973 and the buildings, while well maintained, need to be rebuilt. The buildings are literally falling apart! The architects would like to preserve the playgrounds that all the buildings are centered around and need to know the heights of all seven blocks for the new plan. Find out all the heights of the blocks at Island School and make a scale drawing of the current school. This scale drawing should be a bird's eye view and include all dimensions, including the height. Design your own school, ensuring that you accommodate all faculties and facilities listed below. You may use any format to present your findings.

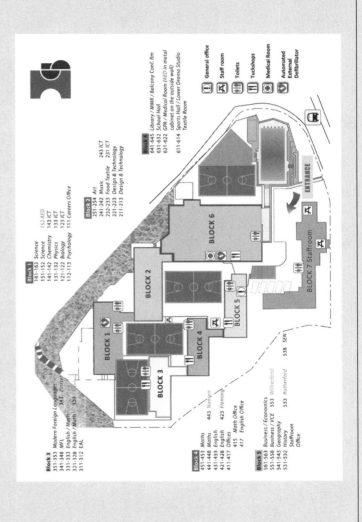

Block 1
161-163 *Science*
151-152 *Science*
141-142 *Chemistry*
131-132 *Physics*
121-122 *Biology*
112-113 *Psychology*

152 AED
143 ICT
133 ICT
123 ICT
111 Careers Office

Block 2
251-254 *Art*
241-242 *Music*
232-233 *Food Textile*
221-223 *Design & Technology*
211-213 *Design & Technology*

243 ICT
231 ICT

Block 3
351-353 *Modern Foreign Languages:*
341-348 *MFL*
331-333 *English / Math*
321-328 *English / Math*
311-312 *EAL*

343 Emson
341 Emson
351

Block 4
451-453 *Maths*
441-448 *Maths*
431-433 *English*
421-428 *English*
411-417 *Offices*
415 *Math Office*
417 *English Office*

443 Nansen
423 Fleming

Block 5
561-563 *Business / Economics*
551-558 *Business / VCE*
541-543 *Geography*
531-532 *History*
Staffroom
Office

553 Wilberforst
533 Rutherford
538 SEN

Block 6
641-645 *Library / MMR / Balcony Conf. Rm*
631-632 *School Hall*
621-622 *GPA / Medical Room (AED) in metal cabinet on the outside wall)*
611-614 *Sports Hall / Lower Drama Studio*
Textile Room

General office
Staff room
Toilets
Tuckshops
Medical Room
Automated External Defibrillator

BLOCK 1
BLOCK 2
BLOCK 3
BLOCK 4
BLOCK 5
BLOCK 6
BLOCK 7 Staffroom
ENTRANCE

FIGURE 6.8: THE TRIGONOMETRIC RATIOS OF SIMILAR RIGHT-ANGLED TRIANGLES

Similar Right-Angled Triangles

Angle Size: _____

Triangle	Opposite Length	Adjacent Length	Hypotenuse Length	$\frac{OPP}{HYP}$	$\frac{ADJ}{HYP}$	$\frac{OPP}{ADJ}$
A						
B						
C						
D						
E						
Mean						

What do you notice about the last three columns?

If you changed the angle in the similar triangles, would you notice the same thing?

What generalization can you make about the ratios in similar right-angled triangles?

Similar Right-Angled Triangles

How many similar triangles are there below?

Explain why the right-angled triangles below are similar.

Label all three sides of the similar right-angled triangles and measure the given angle in the diagram.

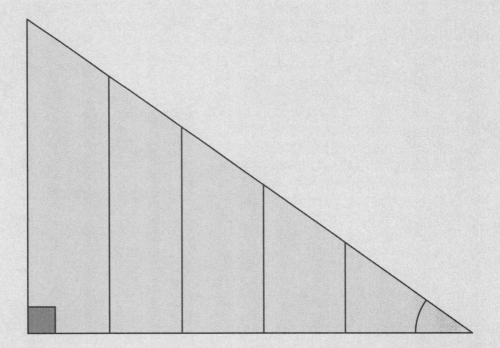

Use the recording sheet and measure the length of the sides of each triangle in the above diagram.

FIGURE 6.9: HOW TO USE TRIGONOMETRY TO MEASURE THE HEIGHTS OF BUILDINGS

How Tall is Block 5?

You will need:

Clinometer

Trundle wheel

Scientific calculator

Clipboard

Pen

Marker

Partner

Island School illustration by Jordan Wathall

1. Measure 20 meters from the base of Block 5 along flat ground and mark the position with a marker.
2. Stand or kneel at the marker and aim the clinometer at the top of Block 5. Release the trigger and ask your partner to read the angle of inclination. Record this overleaf.
3. Ask your partner to measure the height of your eye above the ground to the nearest centimeter. Record this result overleaf.
4. Look at the right-angled triangles ABC and, using the words hypotenuse, opposite, or adjacent, complete the following:

AB is called the _____

Block 5

BC is called the _____

Which trigonometric ratio uses these two sides?

Now use trigonometry to find AB.

The height of block 5 is: =   (AB + my eye height CE) meters
                              =
                              =

Table of Results

Name	Angle	Height below eye	AB	Height of Block 5

Questions:

1. Did everyone get the same result for the height of Block 5? Why or why not?

2. Calculate the mean, median, mode, and range of your results for the height of Block 5. Which is the best average to use for the height of Block 5?

The Frayer Model

The Frayer model (Frayer, Frederick, & Klausmeier, 1969) is a widely used graphic organizer that requires learners to understand not only definitions but also more complex key concepts. Any concept may be analyzed according to certain attributes or facets. Facets of any concept are drawn from students by asking the following questions:

- What is the definition of…?
- List the characteristics of…
- What are some examples of …?
- What are some non-examples of …?

This model promotes a higher level of thinking by asking students to analyze and think of examples and non-examples to assess deep understanding of the mathematical concept. There are several approaches when using the Frayer model with students. Teachers can provide students with a word at the center of the graphic organizer and allow students to fill in the four quadrants. Alternatively, learners may be provided with information for the four quadrants and work out the word, or concept, in the center.

Figure 6.10 is an example of a template utilizing the four facets of the Frayer model.

Figure 6.11 is an example of how the Frayer model could be used for polygons.

 For an additional Frayer model template, please visit the companion website Figure M6.5.

Concept Attainment Cards

Page Keeley and Cheryl Rose Tobey (2011) outline an adaption of the Frayer model by looking at two facets: examples and non-examples. This type of prompt assesses conceptual understanding in a similar way to the Frayer model.

An example of how to use this would be to ask students to provide examples and non-examples of rational numbers. Figure 6.12 illustrates a prompt card that students could fill in individually or in pairs.

Agree, Disagree, and Depends

This is a kinesthetic activity I use a lot if I need quick feedback about the understanding of a particular concept. Three signs—Agree, Disagree, and Depends—are posted on three different classroom walls, and students choose and stand next to a sign depending on the statement presented. Each group is given time to discuss their reasons and justify why they chose the Agree, Disagree, or Depends sign, and

FIGURE 6.10: THE FRAYER MODEL TEMPLATE

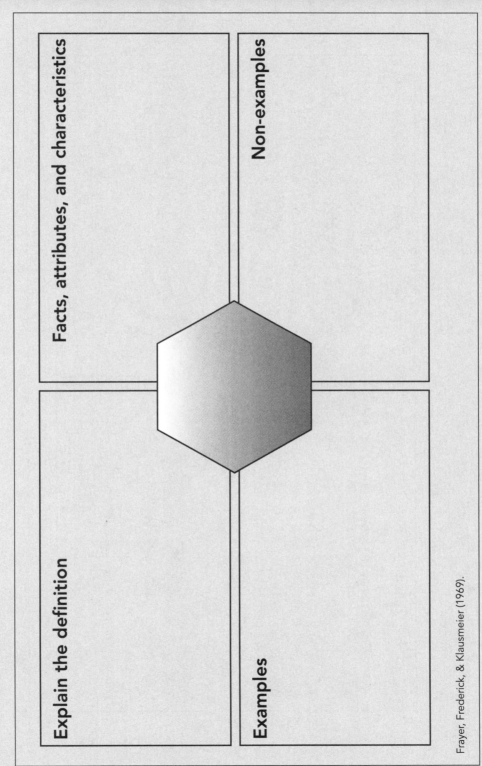

Explain the definition

Facts, attributes, and characteristics

Examples

Non-examples

Frayer, Frederick, & Klausmeier (1969).

173

FIGURE 6.11: EXAMPLE OF THE FRAVER MODEL FOR POLYGONS

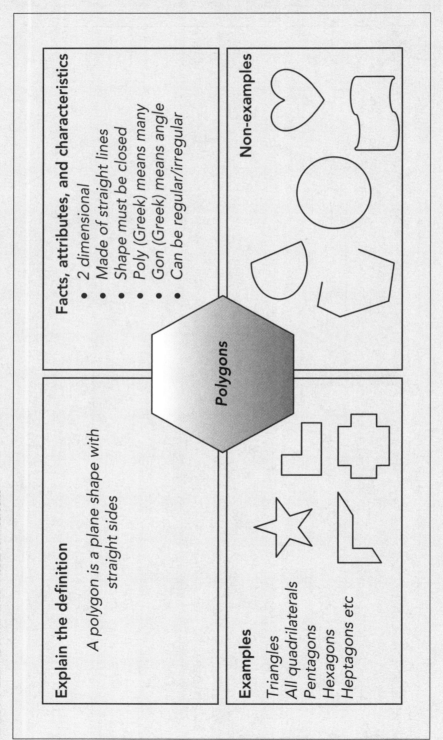

Explain the definition

A polygon is a plane shape with straight sides

Examples

Triangles
All quadrilaterals
Pentagons
Hexagons
Heptagons etc

Facts, attributes, and characteristics

- 2 dimensional
- Made of straight lines
- Shape must be closed
- Poly (Greek) means many
- Gon (Greek) means angle
- Can be regular/irregular

Non-examples

Polygons

FIGURE 6.12: TABLE OF EXAMPLES AND NON-EXAMPLES FOR RATIONAL NUMBERS

Examples of Rational Numbers	Non-Examples of Rational Numbers

feedback is given to the class. The "depends" sign can be adapted to a "don't know" sign. This is a sample statement:

All quadratic equations can be solved using the quadratic formula.

This type of question draws out whether students understand that different tools support the problem-solving process. However, the selection of a method is key to mathematical efficiency and can reveal the beauty in mathematics.

> If students answer *agree,* then they do not understand that the quadratic formula may not be the most efficient method.
>
> If students choose *disagree,* this means they do not understand that actually the quadratic formula can solve all quadratic equations, even though it might not be the most efficient method.
>
> If students choose *depends,* they need to justify why.

Tobey and Arline (2014a) provide excellent probes that could be used for Agree, Disagree, or Depends (see Figure 6.13).

FIGURE 6.13: PROBE FOR AGREE, DISAGREE, OR DEPENDS

Value of the Inequality

If $m > 0$ and $n < 0$ decide whether you

agree, disagree, or think it depends for each question.

1. $m + n < 0$
2. $m - n > 0$
3. $(m)(n) < 0$

Adapted from Tobey and Arline (2014a).

Zero, One, Two, or Three

Some students are inhibited about expressing their understanding in front of the class. A more discreet strategy I use is zero, one, two, or three. During a lesson, students write in their books either zero, one, two, or three (or this could be how many fingers up in front of the chest) to indicate their level of understanding.

0 = didn't understand

1 = understood a bit

2 = understood most

3 = completely understood

The teacher can easily go around the classroom and support students accordingly. I also use this strategy with homework and ask students to write zero, one, two, or three at the top of homework before it is handed in. This allows me to see which students need mini workshops and reinforcement of the concepts covered before collecting the homework.

The Use of Multiple Choice Questions

Multiple choice questions are often used as a summative assessment tool; however, they can be used as effectively, if not more so, as diagnostic and formative assessment tools to check for student understanding during a unit of work. It is important that students are encouraged to justify their choice to reflect their thinking and understanding.

Here is an example of a multiple choice question:

> If $P(A)$ is the probability that an event will occur, which of the following must be false?
> a. $P(A) = 1$
> b. $P(A) = \dfrac{1}{2}$
> c. $P(A) = \dfrac{1}{3}$
> d. $P(A) = -1$

Here students need to understand that probabilities cannot be represented by negative values and that the probability of an event could be 100% (choice d).

Assessing and Developing Core Transdisciplinary Skills

The development of the approaches to learning skills in our students is crucial to preparing our students for the future. Approaches to learning include the following generic skills

- Cognitive skills, such as thinking skills;
- Metacognitive skills, which involve the awareness and ability to understand and control one's cognitive processes;
- Affective skills, which include behavior and emotional management.

Surveys conducted with corporations list the top four generic skills as critical thinking and problem solving, information technology, teamwork and collaboration, and creativity and innovation (Casner-Lotto & Benner, 2006).

The International Baccalaureate organization has introduced integrating core transdisciplinary principles in all their programs (Primary Years Program, Middle

Years Program, Diploma Program, and IB Career-Related Program) called the "Approaches to Teaching and Learning." The purpose of approaches to learning is to highlight the importance of the development of skills in our students. The five categories of skills in the approaches to learning are as follows:

1.   Communication
2.   Social
3.   Self-Management
4.   Research
5.   Thinking

The approaches to teaching are as follows:

1.   Use of inquiry
2.   Emphasis on conceptual understanding
3.   Developed in local and global contexts
4.   A focus on collaboration
5.   Allows for differentiation
6.   Informed by summative and formative assessment

Lance King (2014) has been instrumental in the development of the ten–cluster model of Approaches to Learning skills for the new MYP (Next Chapter) level of the International Baccalaureate. The development of these skills is vital in preparing our students for future success.

Figure 6.14 outlines a self-assessment rubric developed by Lance King based on Dreyfus and Dreyfus (2000) and Berliner's (2004) work on assessment of skills. It can be seen that this rubric is in line with both Maslow's (1987) level of learning and Stephenson's (1998) development of capability.

Assessing the Developing Concept-Based Student

How do we know whether our students are developing into concept-based students? Lois Lanning has created a rubric that contains a checklist of three areas for the developing concept-based student: task commitment, synergistic thinking, and depth of understanding (Erickson & Lanning, 2014). The rubric outlines the progression from novice to mastery level in all three of these areas. Encouraging students to self-assess their position in the rubric will give learners a better understanding of how to develop concept-based learning.

It is important to highlight the mastery levels for task commitment, synergistic thinking, and depth of understanding so students understand the goals for developing into a concept-based student.

FIGURE 6.14: ASSESSING APPROACHES TO LEARNING SKILLS

	Approaches to Learning Skills Proficiency			
	Level 1 The Novice Observation	Level 2 The Learner Emulation	Level 3 The Practitioner Demonstration	Level 4 The Expert Self-Regulation
Dreyfus & Dreyfus (2000)	Observes others performing tasks and using the skill	Copies others' performance of the skill	Can demonstrate the skills on demand	Can perform the skill without thinking through the process first
Berliner (2004)	Gains an understanding of how the skill operates and what the distinguishing characteristics of the skill are Gathers procedural information about the performance of the skill, asks questions to clarify procedure Errors are frequent High levels of scaffolding from teacher needed: explanations, training, structural support	Works through the skill in a step-by-step fashion, seeks clarification for correctness of performance Consolidation of learning is occurring through experience Is very conscious of performing the skills and correcting errors with deliberation Medium level of scaffolding needed: correcting poor performance, answering questions	Flexibility of skill use in different contexts is developing Automaticity is developing Errors are corrected quickly Minimal teacher scaffolding required: setting directions, goals, assessable outcomes	Can teach others the skill Automaticity is established High level of performance occurs Any errors are corrected automatically No teacher scaffolding needed
Stephenson (1998)	Is not performing the skill	Performs skill only with known content in known context	Can perform skills either with different content or different context	Can use skill with unfamiliar content in unfamiliar context
Maslow (1987)	Unconscious incompetence	Conscious incompetence	Conscious competence	Unconscious competence

© Lance King (2014) ATL Skills

The mastery levels are detailed as follows:

Mastery level for task commitment includes qualities such as being able to

- Independently seek to extend the assigned work
- Express enthusiasm and genuine interest in the challenge of the work
- Demonstrate self-discipline and persistence with longer term and more complex work
- Value and seek collaboration to complete task
- Show an equal interest in the learning process as producing a quality product

Mastery level for synergistic thinking includes a student being able to

- Transform examples and ideas into higher level generalizations
- Articulate generalizations and clearly support them with multiple examples
- Independently link factual examples to related concepts and generalizations
- Independently provide examples and non-examples of concepts
- Demonstrate thought processes that show unique and inventive connections between generalizations, concepts, examples, and product

Mastery level for depth of understanding includes students being able to

- Systematically and methodically explain understanding supported by relevant evidence
- Create novel products
- Create new knowledge by extending ideas, experimenting with product guidelines, applying different learning strategies
- Show the ability to prioritize and critique the relevance of examples
- Transfer generalization across multiple more complex examples
- Use conceptual language confidently and accurately
- Independently provide examples and non-examples of concepts

Figure 6.15 shows the descriptors for each level on the rubric, from novice to emerging to mastery level. This rubric allows students to identify where they are and where they need to go in order to develop into a concept-based student.

Self-Assessments

Self-assessment strategies can develop metacognitive abilities, which involve students being able to self-monitor and reflect. Research supports that developing students'

FIGURE 6.15: RUBRIC: THE DEVELOPING CONCEPT-BASED STUDENT

	Novice	Emerging	Master
Task Commitment	• Views work as compliance versus ownership of task • Tasks are seen as separate and unrelated • May become impatient and easily frustrated • Stays within the guidelines of the assignment	• Sees the learning relevance of assigned work • Stays clearly focused and self-regulated throughout the task • Is willing to persevere through challenges that are within reach • Accepts responsibility for work • Actively participates in collaborative groups • May incorporate new directions or approaches to an assignment	• Independently seeks to extend the assigned work • Expresses enthusiasm and genuine interest in the challenge of the work • Demonstrates self-discipline and persistence with longer term and more complex work • Values and seeks collaboration to complete task • Is as interested in the learning process as producing a quality product
Synergistic Thinking	• Recognizes connections and patterns among ideas or solutions when they are pointed out • Inconsistently ties concepts to the presented examples • Makes generalizations that are overly simplified	• Synthesizes information into a coherent generalization • Supports generalization with at least two accurate examples • Links factual examples to presented concepts	• Transforms examples and ideas into higher level generalizations • Articulates generalizations and clearly supports them with multiple examples • Independently links factual examples to related concepts and generalizations • Independently provides examples and non-examples of concepts • Thought processes show unique and inventive connections between generalizations, concepts, examples, and product

(Continued)

181

	Novice	Emerging	Master
Depth of Understanding	• Needs support with attempts to explain learning targets, thinking, and generalizations • Generates relevant questions but most are still at the factual level	• Clearly explains personal position and acknowledges other perspectives • Produces thoughtful, detailed products • Formulates relevant questions • Incorporates prior learning into present context • Transfers generalizations across a few concrete examples • Expands use of conceptual language in explanations • Recognizes non-examples of concepts	• Systematically and methodically explains understanding supported by relevant evidence • Creates novel products • Creates new knowledge by extending ideas, experimenting with product guidelines, applying different learning strategies • Shows the ability to prioritize and critique the relevance of examples • Transfers generalization across multiple, more complex examples • Uses conceptual language confidently and accurately • Independently provides examples and non-examples of concepts

Transitioning to Concept-Based Curriculum and Instruction, Corwin Press Publishers, Thousand Oaks, CA p. 77.

FIGURE 6.16: EXAMPLE OF A SELF-ASSESSMENT WORKSHEET

Evaluation: Attitude to Learning	Grade: *Topic:*

Reflect on your attitude to learning by completing the following self-evaluation, putting a number 0, 1, 2, or 3 in the boxes.

0 = not at all

1 = a bit

2 = mostly

3 = always

	Student		Teacher
1. I complete all homework, class work tasks, and all worksheets.			
2. I check all work and answers carefully and learn from mistakes.			
3. All of my work is well organized and well presented.			
4. My diagrams, graphs, and tables are clearly and accurately presented.			
5. I have reviewed the key concepts by making study notes and underlining or highlighting key ideas.			
6. My written answers are given in detail and communication is clear, showing all methodologies employed.			
7. I collaborate well and possess good teamwork skills.			
8. I am willing to contribute to class discussions and seek assistance when necessary.			
9. I have a positive attitude to study and I am well motivated.			
10. I bring all necessary equipment to lessons and I am prepared for learning in my lessons.			

Teacher Evaluation

Your teacher will highlight a few of these skills as targets for future development.

Teacher feedback:

(Continued)

Name: _____

Year 10 Mid-Unit Geometry Assessment

Dedicated Improvement and Reflection Time (D.I.R.T)
Self-Assess and Reflect

Area of Focus Dedicated Improvement and Reflection Time	Section Question	Your score	Do you understand? Explain
You understand how to express the Pythagorean theorem in words, give examples, list characteristics, and also provide non-examples of the theorem.	A		
Know how to substitute numbers into the Pythagorean theorem	B1		
Know how to find the area and perimeter of right-angled triangles	B2		
Be able to add up middle side and longest sides in a right-angled triangle	B3		
Understand how to find a rule connecting the shortest side and sum of the other two sides in a right-angled triangle	B4		
Understand how to find a rule that connects the perimeter and area of right-angled triangles	B5		
You understand how to express the ambiguous case of the sine rule in words, giving examples and why there are sometimes two triangles that can be drawn for the Donkey case.	C		

Areas I need to revise are

I am pleased with

Student's signature:_____

Teacher's Signature:_____   Date:_____

ability to become more self-aware of and engaged in his or her own learning helps student achievement. Students can be encouraged to reflect through the vehicle of self-assessments.

D.I.R.T. refers to Dedicated Improvement and Reflection Time and encourages students to process and reflect on feedback from the teacher and on their own learning (Beere, 2013). Students understand the purpose of feedback given to them if they have been given opportunities to reflect and respond.

Figure 6.16 shows an example of a self-assessment task with time allocated for D.I.R.T.

Traditional textbooks focus on the repetition of mathematical skills and strategies and deal tenuously with the conceptual understandings of the content through deductive approaches. Traditional assessment is commonly grounded in closed summative approaches that rely on memorization of facts and skills.

A variety of diagnostic and formative assessment strategies give teachers and learners valuable feedback to inform future instruction during the learning process. Tracking students' starting and finishing points identifies where a student makes progress in learning—the essence of formative assessment. The main purpose of any form of assessment is to evaluate the effectiveness of the curriculum and instruction.

Chapter Summary

This chapter provided several assessment strategies, with examples, to engage students and assess their conceptual understanding. Assessment helps to inform instruction and student progress.

A number of assessment strategies were discussed:

Assessments with conceptual depth, such as open inquiry tasks and open-ended questions, provide teachers with opportunities to improve teaching and learning. Open inquiry tasks are student directed and develop students' independence to be lifelong learners. Inquiry-based assessment asks students to describe the conceptual understandings involved to reflect understanding and purpose for what they are learning.

Visible thinking routines (from Harvard University's Project Zero) are strategies to make student thinking visible. These routines allow learners to express and communicate thoughts and ideas.

Quality performance tasks assess students' knowledge (factual content), what they can do (processes and skills), and conceptual understanding (generalizations or

(Continued)

(Continued)

principles), all through an engaging scenario. RAFTS is a planning tool to help design a performance task by looking at role, audience, format, topic, and the use of strong verbs. The structure of the task planner—what, why, how—promotes the development of quality performance tasks (Erickson, 2007).

Frayer's model is another widely used assessment tool, which provides opportunities for learners to identify different attributes of a concept by looking at examples as well as non-examples to show conceptual understanding.

Agree, Disagree, and Depends is a kinesthetic activity that allows students to move around the room while promoting discussion. A statement is posed and students are asked to stand next to the sign that represents either agree, disagree, or depends. Students are given opportunities to discuss, justify, and present their points of view.

The International Baccalaureate organization has developed the Approaches to Teaching and Learning, which highlight the importance of developing transdisciplinary skills such as communication, social, self-management, research, and thinking. Lois Lanning's rubric for developing the concept-based student is included in this chapter and provides a roadmap for increasing conceptual thinking and performance.

The next chapter will look at the effects of technology supporting conceptual understanding in mathematics.

Discussion Questions

1. Choose three of the assessment strategies listed in this chapter. When, why, and how would you use each in your own classroom practice?

2. What are some advantages of visible thinking routines?

3. How would you use a visible thinking routine?

4. Are there other graphic organizers you use that highlight students' conceptual understanding?

5. What math topics lend themselves well to performance assessment tasks?

6. How do you develop core skills such as communication, self-management, and research skills in the math classroom? What might you do in the future to improve your practice in developing these skills?

7. How might you use the rubrics in this chapter for developing the concept-based student in your own classroom context?

CHAPTER 7

How Do I Integrate Technology to Foster Conceptual Understanding?

When I first started teaching, there was no World Wide Web or even computers. I would spend hours preparing overhead projector laminates, which was a lot more efficient than writing on the blackboard. At the time, I would not have been able to imagine how my teaching would completely transform due to the technological advances of the last 30 years.

I have always been an enthusiastic and avid user of technology. I enjoy experimenting with new technological tools to see whether they impact and enhance learning. I have observed lessons where teachers lecture, didactically, from PowerPoint slides and demonstrate wonderful applets and animations from mathematics websites. However, replacing writing on a blackboard with PowerPoint slides is utilizing technology as a mere substitution tool. Showing your students a wonderful applet can help students understand a concept on some level, but it does not encourage students to undertake the inquiry or thinking process to make deeper connections between the concepts. Regarding the effective integration of technology, my motto has always been "it's not *what* you use, it's *how* you use it."

> "e-Learning" is a misnomer – it's mostly just "e-Teaching." For any teaching to reliably and consistently produce the results we want, we still have a lot to learn about learning. (Prensky, 2002)

Effective integration of technology should enhance conceptual understanding and learning for our students. Over the last three decades, the digital revolution has had a profound effect on education and how students learn. Some classrooms around the world are 1:1 (one student, one laptop), giving students access to a vast range of digital tools. Our students are part of the e-generation; they are digital natives who do not know a world without the World Wide Web, laptops, or iPhones.

Puentendura (2013) has introduced a technology integration framework called **SAMR** (substitution, augmentation, modification, and redefinition), which describes the hierarchical levels of technology use. The lower level starts with the technology being used as a direct substitution tool, such as reading from a PowerPoint slide instead of writing notes on a blackboard. From substitution there is a progression to three other levels: augmentation, modification, and redefinition. This model can be used to describe the levels teachers follow as they adopt technology. Figure 7.1 summarizes each level and shows examples in the mathematics classroom.

When utilizing digital tools in the classroom, one would hope that technology is used at the augmentation level, at a minimum, to enhance learning in our students.

TPACK is a popular framework that explains how teachers' understanding of educational technologies and their pedagogical and content knowledge interact with one another to produce effective teaching with technology. All teachers have in their toolbox content knowledge and pedagogical knowledge. This is known as CPK (content pedagogical knowledge, from the work of Lee Shulman, 1986). A third component—technological knowledge—must be added to this toolbox to keep up with the 21st century.

TPACK outlines three main knowledge areas for effective technology integration:

1. Content knowledge (CK) represents the subject knowledge we are teaching.
2. Pedagogical knowledge (PK) refers to the approaches to teaching, such as inquiry-based, problem-based, or collaborative learning, to name a few.
3. Technological knowledge (TK) refers to the technological tools that we use to make the content more accessible to students.

Once we understand each of these knowledge areas, we can look at how they interact with adjacent areas (see Figure 7.2). The relationships among the three knowledge areas account for the wide variations in the quality of educational technology integration.

Koehler and Mishra (2009) have completed extensive work constructing the TPACK model: "The TPACK framework for teacher knowledge is described . . . as a complex interaction among three bodies of knowledge: content, pedagogy, and technology. The interaction of these bodies of knowledge, both theoretically and in

Level		Description	Examples When Learning Mathematics
Substitution	Enhancement	Technology tools act as a direct substitution with no functional change.	Students type up projects instead of handwriting. Teachers use PowerPoint slides instead of writing on the board. Students look up the definitions of math words on the Web instead of using a dictionary.
Augmentation		Technology tools act as a direct substitution with functional improvement.	Students collect survey results using online survey tools or Google forms. Students practice questions on online websites.
Modification	Transformation	Technology allows for significant task redesign.	Choose one of the many proofs for the Pythagorean theorem and produce a video to explain this proof using animation. Producing videos for a flipped classroom. Use Google Docs or Presentation as a collaboration tool. Use various apps for formative feedback.
Redefinition		Technology tools allow for the creation of new tasks previously unconceivable.	In groups, create a website with different pages including multimedia such as audio recordings and videos to demonstrate conceptual understanding. Use Twitter to get feedback or collect data from around the world. Use Skype to collect data by conducting interviews around the world. Use Google Earth, an interactive three-dimensional environment that allows the study of concepts such as scientific notation, estimation, and time zone problems. Flat classroom projects—this is a global collaborative project inspired by Thomas Friedman's book *The World is Flat,* which fosters communication and interaction between students and teachers across the world.

Adapted from Puentendura (2013).

practice, produces the types of flexible knowledge needed to successfully integrate technology use into teaching" (p. 60).

For effective technology integration, the three knowledge areas interact as illustrated in the Venn diagram in Figure 7.2.

FIGURE 7.2: THE TPACK MODEL FRAMEWORK

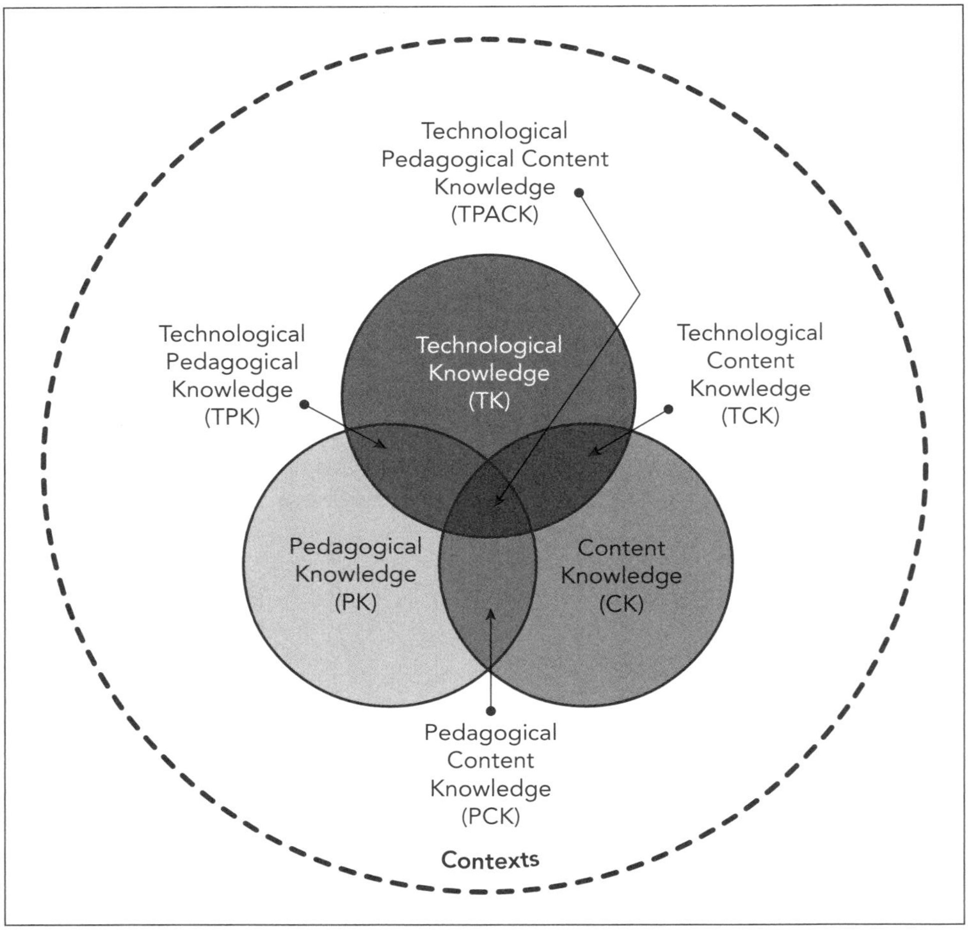

Reproduced by permission of the publisher, © 2012 by tpack.org

Notice in the Venn diagram the overlap of the different adjacent areas of knowledge:

Technological pedagogical knowledge (TPK): This is an understanding of how teaching and learning can change when particular technologies are employed.

Technological content knowledge (TCK): Technology has made a huge impact on our content knowledge and resulted in advancement in many fields, such as medicine, physics, and history. The understanding of how technology has impacted our content knowledge is critical to developing appropriate technological tools for education.

Pedagogical content knowledge (PCK): This is an understanding of our approaches to teaching and learning the content of our discipline area. This knowledge exists in teacher's toolboxes.

At the heart of good teaching is the intersection of all three areas: technological pedagogical content knowledge (TPACK). This is an understanding that knowledge emerges beyond the three individual components. When these three components interact with each other, they represent optimal conditions for successful technology integration. Koehler, Mishra, and Cain (2013) describe the TPACK framework as one that "seeks to assist the development of better techniques for discovering and describing how technology-related professional knowledge is implemented and instantiated in practice. By better describing the types of knowledge teachers need (in the form of content, pedagogy, technology, contexts, and their interactions), educators are in a better position to understand the variance in levels of technology integration that occurs" (p.18).

This framework can support the unit planning process, as it shows how technology can enhance and be embedded in any subject.

Here is an example of how to use this framework when planning a lesson. This learning experience can be completed at the beginning or end of a unit:

Content: Students collaborate to show what they know about a topic, such as mathematical modelling.

Pedagogy: Use a concept map or mind map, which may also connect with skills development such as critical thinking, communication, collaborating, and creativity.

Technology: Use a digital mind mapping tool such as Mind42 (http://mind42.com) or Padlet (http://padlet.com) to create the concept map.

In the next few sections, I will outline several digital tools that promote inquiry and conceptual understanding in mathematics.

Mathematics Graphing Software and Graphical Display Calculators

Mathematics graphing software and graphical display calculators have allowed students to focus less on computation and more on understanding the concepts of a particular topic. Figures 7.3 and 7.4 are examples of inquiry activities for which any graphing software or graphical display calculator can be utilized to support students' understanding of the concept of transformations and circle theorems.

Figure 7.3 guides students through inductive inquiry to understand the effects of the parameter k and how it can transform a graph of a function. An inductive approach is adopted whereby students look at specific examples and then are asked to generalize. Without this digital graphing tool, drawing individual graphs would be time consuming and laborious. This would also result in an overemphasis on tedious drawing rather than a focus on the concepts that derive from the content.

Technology can support conceptual understanding by presenting information instantly, enabling more efficient pattern spotting. Graphing software can also help students to visually represent more complicated functions for further analysis. One I would highly recommend is Douglas Butler's "Autograph" (http://www.auto graph-maths.com). As a mathematics educator, he has designed a dynamic platform for mathematics software to help students understand the concepts in many topics. Freely available GeoGebra (https://www.geogebra.org) is also useful as a learning and teaching tool that teachers and students can use to build in-house applets. Desmos (https://www.desmos.com) is another freely available online graphing software tool with numerous ready-built activities for the classroom.

Figure 7.4 illustrates how using graphing software helps students to discover relationships between arcs and angles in various circle theorems. Students build applets and look at specific values for the angles subtended and then are asked to generalize.

Figures 7.3 and 7.4 illustrate how graphing software can be utilized to support the inductive approach to learning new concepts. Students are asked to make generalizations and pattern seek during both of these learning experiences.

Flipped Classroom

The digital age has given birth to the "flipped classroom" approach, which places the responsibility for learning on the student, with the teacher acting as a facilitator rather than an information disseminator. A typical flipped classroom is organized as follows: Students learn the content of the subject through instructional videos outside the classroom. Lessons with teachers focus on the problems and challenges students encounter with the material being learned.

Hamdan, McKnight, McKnight, and Arfstrom (2013), in an article titled *A Review of Flipped Learning* discusses the success that Clintondale High School in Michigan has experienced with the flipped classroom approach. Clintondale High School implemented flipped classrooms in all grades and all subjects in 2011. After the introduction of this model, Clintondale's failure rate dropped from 30% to 10%, and there were remarkable improvements in achievements in all subject areas. Students reportedly responded well to the flipped learning model.

FIGURE 7.3: USING GRAPHING SOFTWARE

Transformation of Curves

Use your graphical display calculator or any graphing software for the following activity and let $y = f(x)$

Sketch the following on the same axes $y = x^2$, $y = x^2 + 2$, $y = x^2 - 4$	Sketch the following on the same axes $y = x^2$, $y = 2x^2$, $y = -3x^2$
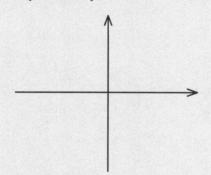	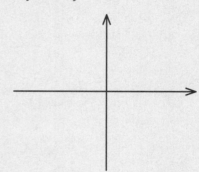
Describe the effect of k for $y = f(x) + k$	Describe the effect of k for $y = kf(x)$
Sketch the following on the same axes $y = x^2$, $y = (x - 1)^2$, $y = (x + 2)^2$	Sketch the following on the same axes $y = x^2$, $y = (2x)^2$, $y = (4x)^2$
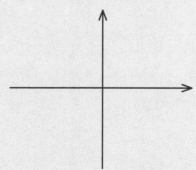	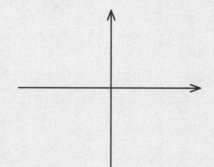
Describe the effect of k for $y = f(x + k)$	Describe the effect of k for $y = f(kx)$

Sketch the following on the same axes

$f(x) = e^x + \sin x$

(i) $y = f(x) + k$

(ii) $y = f(x + k)$

(iii) $y = kf(x)$

(iv) $y = f(kx)$

Explain each transformation on your sketch:

Summarizing Your Findings

Transformation	Generalization
$y = f(x) + k$	
$y = f(x + k)$	
$y = kf(x)$	Special Case: $k < 0$
$y = f(kx)$	

 For a completed version of Figure 7.3, please visit the companion website.

FIGURE 7.4: USING GRAPHING SOFTWARE FOR CIRCLE THEOREMS

Circle Theorems

Using any graphing software package, draw a circle of any size.

1. Reproduce this diagram by drawing a diameter and two chords to form a triangle.
    Measure the angle subtended at the circumference.
    Move all three points that are on the circumference of your circle.
    What do you notice about the angle?
    What generalization can you make about the angle subtended in a semi-circle?

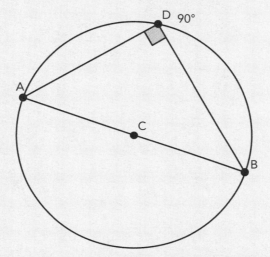

2. Now draw this diagram below by drawing a circle, a radius and a tangent.
    Measure the angle between the radius and the tangent.
    Move the point on the circumference. What do you notice about the angle?

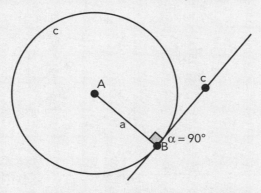

What generalization can you make about the angle formed by the radius and tangent in a circle?

3. Reproduce the circle below where A is the center and D is on the circumference.

Measure the angles at the center and at the circumference (i.e., angle CAB and angle CDB).

Move the point D around the circumference and also change the size of the circle. What do you notice about the angles subtended at the center and the angle subtended from the arc to the circumference (i.e., angle CAB and angle CDB)?

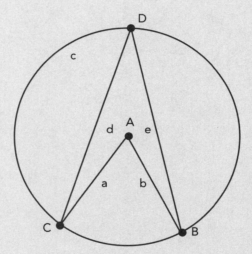

What generalization can you make about the angle subtended from the same arc at the center of the circle and the circumference?

(Continued)

4. Reproduce the following circle, where A is the center of the circle and angles are subtended from the same arc to the circumference of the circle.

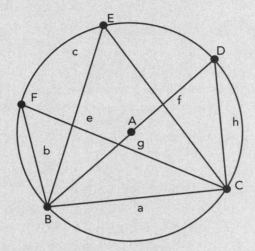

Move the points D, E, and F on the circumference of your circle and change the size of your circle. What do you notice about these angles subtended from the same arc?

What generalization can you make about the angle subtended from the same arc to the circumference of the circle?

Summarize your findings

	Generalization
Angle subtended at the circumference	
Angle between tangent and radius	
Angle at center, angle at circumference	
Angles from the same arc	

In a flipped classroom, also known as the reverse or inverted model of instruction, teachers concentrate on what students do not know rather than spending time in lessons on what students already understand. Often, misunderstandings and misconceptions are dealt with in a more efficient way in the flipped classroom model.

Figure 7.5 is an example of how to use the flipped classroom approach by watching various videos on complex numbers on YouTube for homework (e.g., Khan Academy, 2014). Class time is spent on assessing students' understanding of this topic if they can answer the questions in Figure 7.5. Class time is also spent on clearing up misconceptions and focusing on what students do not understand rather than what they do understand.

Multimedia Projects

A few years ago, I had a student, Jacob (not his real name), who was an aspiring filmmaker. I had five students in the class who all struggled with math throughout their school life. They were all enrolled in a compulsory mathematics course called the IB Mathematical Studies (SL) course. This course is designed for students who pursue further studies in disciplines unrelated to mathematics.

These students joined my class with a history of math anxiety and low motivation for the subject. In order to captivate them and utilize their talents, we spent an entire year making videos on the various topics of the course. Sometimes these were instructional videos, with the group taking different roles in explaining various concepts and examples. Sometimes they wrote songs to help them remember certain key facts or processes, and other times they just made advertisements for a particular topic in math by highlighting the main ideas and real-life applications. Jacob acted as the director, which not only utilized his filmmaking talents but also gave him an incentive to engage with and access the math concepts. This worked extremely well for this group, and now I extend multimedia projects to include creating online brochures, news reports, and interactive websites.

These types of activities are always completed in pairs or groups so collaborative and communication skills are also developed. Giving this class the option of using multimedia increased their motivation and enthusiasm for math. One student said to me, "I always dreaded going to math lessons until this year." All students passed the course with flying colors, without having to complete any extra work at home. The video-making process allowed my students to discuss, plan, and understand the concepts involved in topics on a deeper cognitive level due to the high level of engagement.

FIGURE 7.5: FLIPPED CLASSROOM LESSON ON COMPLEX NUMBERS

Imaginary and Complex Numbers

Solve the following equation:

$$x^2 - 4x + 13 = 0$$

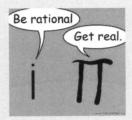

The square root of -1 is defined as *i* (for "imaginary"!). Write down the relationship between 1 and *i*

You found the question above had no real solutions. Using $\sqrt{-1} = i$ rewrite the solution to the quadratic above.

Notice that you have a real part and an imaginary part to your solution. What do you notice about your solution?

For a real quadratic equation with $\Delta < 0$, if $a + bi$ is a complex root, then _____ is also a root.

A complex number, z, is:

Skills Practice

Solve for x:

$x^2 - 10x + 29 = 0$

$2x + \dfrac{1}{x} = 1$

The Number System

Define the following symbols:

$\mathbb{N}$

$\mathbb{Z}$

$\mathbb{Q}$

$\mathbb{R}$

Draw a Venn diagram to show the relationship between: $\mathbb{N}$, $\mathbb{Z}$, $\mathbb{Q}$, $\mathbb{R}$

How do you include irrational, imaginary, and complex numbers in your Venn diagram?

 For a completed version of Figure 7.5, please visit the companion website.

Collaboration Tools: Google Applications

A Google Doc is an online, live word processor that allows collaborators to create and edit a document in real time. Google Slides and Google Sheets work in a similar fashion to Microsoft PowerPoint and Excel, respectively. These live Google applications promote collaboration, creating a social learning environment and ensuring all students' ideas are collated in a real-time format. Here are some examples of group research projects to construct learning using Google applications in the classroom:

Google Slides

Below is an example of a collaborative classroom project using Google Slides.

In your groups, find out how to complete the square when solving a quadratic equation, and explain the method using Google Slides.

Ensure that every group member has a specific role and task. Make sure you explain the method of completing the square when the leading coefficient is 1 and also when it is not 1. Answer the question, "Why is this method called completing the square?"

Each member of your group must provide a specific example and include a diagram. Groups will be asked to present to the class.

Google Docs

Similar to Google Slides, Google Docs encourages collaboration both in and out of the classroom and is an efficient way to collect ideas from your students. Before final examinations, I ask students to form groups and choose topics they need to review. They then work in groups to write study notes, and at the end of this process, the review booklet is printed and bound. Access to this Google Doc allows me to check everyone's understanding and modify any misconceptions through discussions with individual students or groups. Here is an example of the instructions I would give:

There are six units of work that need to be reviewed for your final examinations (Algebra, Statistics, Calculus, Vectors, Functions, and Trigonometry).

Choose a unit that you think you need to review and form a group with other students in the class.

Using a Google Doc, write study notes on the key concepts and areas that you need to know and understand. Include examples, diagrams (such as concept maps), and clear explanations.

This work will be produced as a bound review booklet for the class to share.

Google Forms from Google Sheets

Google Forms can be found in the Google Sheets and is a very efficient way for students and teachers to collect survey data. Data on a Google Form is collected on a spreadsheet and allows for efficient data analysis. The benefit of using a Google Form is that data can be collected on a global scale, so students can study the differences between cultures, governments, or any systems.

Here is an example of a task that could involve the collection of data from schools in Hong Kong and the United States.

Do students in Hong Kong play fewer sports than students in the United States?

Design a Google Form with questions that will help you to find out the answer to this question.

Apps on Mobile Devices

Most of my students are fortunate enough to have mobile phones or some form of portable electronic device. These portable devices are able to support well-designed teaching and learning applications (apps). There are a vast number of freely available educational tools that can provide formative feedback, game-based learning environments, and collaborative concept maps.

Collaborative Thoughts and Concept Maps

I often use digital concept maps (sometimes called mind maps) to collect ideas from my students because they include everyone's contributions during and after the lesson. Every idea is valued on a concept map. A concept map is completed for every topic, providing comprehensive review and consolidation notes for students. Concept maps are also excellent for teasing out misconceptions and because of the live nature, misconceptions can be efficiently addressed and amended on the concept map.

Padlet (formerly known as "WallWisher") is free, easy to use, and only requires students to have access to your Padlet address to add their ideas to the wall. This website (http://padlet.com) can be used to collect ideas for a concept map or to make thinking visible in your classroom. For example, I created a Padlet wall (http://padlet.com/jenniferwathall/qv59qrteos) for a Think, Pair, and Share activity about critical thinking.

Another of my walls (http://padlet.com/jenniferwathall/rehh8usohn) shows how Padlet can be used for target setting. The instruction to students at the beginning of the unit was "Choose three critical thinking attributes you would like to develop and explain why."

Another wall (http://padlet.com/MessyMix/GroupA) shows what students (in groups) recapped after a unit on trigonometry.

Another popular free mind mapping tool is Mind42 (http://mind42.com). Sign-up is free and you are also given access to mind maps that others have created.

Exit and Entrance Ticket Apps

The exit and entrance ticket strategy was mentioned in Chapter 6. There are many free apps that will record students' exit and entrance ticket responses for later analysis. This formative feedback strategy allows teachers to assess student progress and plan future areas of focus. Students download a free app on their mobile device, or any device with Internet access, and the website provides you with a class code for students to access your quiz or exit ticket.

Mini Electronic Whiteboard

Another app I highly recommend is one that converts your tablet or iPad to a mini interactive whiteboard, such as Doceri (https://doceri.com) and Educreations (https://www.educreations.com). Rather than focus on the front of the room, students can pass tablets and iPads around the room or within a group to promote collaboration and group problem solving. Chapter 5 discussed the flexible front classroom structure and the benefits of the teacher being immersed in the learning environment. The iPad or tablet can be displayed on monitors around the room wirelessly using an Apple TV, and students will have a digital record of their work for future reference.

Educational "Facebook"

One unique, free website I love to use is Edmodo (https://www.edmodo.com). This is similar to an educational Facebook and allows teachers to organize classes; set homework, quizzes, and quick polls; and upload lessons in a social setting. Students can submit assignments and post messages on the group wall. Students download a free app, which notifies students of alerts from the teachers. No longer will students be e-mailing you about missed work as everything is posted on the class wall. Edmodo is a wonderful organization tool that enables teachers and students to have direct dialogue with their classes. Newly developed Google Classroom works in a similar fashion.

When Not to Use Technology

A Word of Caution

As I mentioned before, keeping the SAMR model in mind, if the digital tool does not enhance learning and is used as a mere substitution tool, then go back to using good old-fashioned markers and poster paper. My famous spaghetti activity is an example that develops students' deep understanding of circular functions and does not require any use of technology.

Figure 7.6 is my famous spaghetti activity that I use in all of my workshops, which helps students understand circular functions and how to generate the graph of a sine function. There are many applets online that demonstrate this concept in 30 seconds; however, the investment in time for this activity (around 80 minutes) allows students to think for themselves and to gain an invaluable understanding of the concept of circular functions and the sine function curve, which can be used to solve more complex problems in trigonometry.

Figure 7.7 is a follow-up activity for sine curves and spaghetti using the unit circle. The unit circle is placed on a grid with a moving arrow, and students form different angles with the arrow to work out the trigonometric ratios.

Figure 7.8 is another activity using raw spaghetti that enables students to discover the relationship between the three sides of a triangle: that the two shorter sides must sum to be greater than the third side. This activity uses the fact that the shortest distance between two points is a line.

The vast advantages of utilizing technology to reinforce conceptual understanding outweigh the opposition to 1:1 programs. If technology is used to enhance learning and conceptual understanding, digital tools can solve complex problems and help students to understand the underlying concepts efficiently and effectively. Remember, importance is not placed on any particular tool but on *how* a digital tool can be used to enhance the understanding of concepts in a particular unit.

FIGURE 7.6: SINE CURVES USING SPAGHETTI

Unit Circle Activity: Sine Curves Using Spaghetti

We will explore the unit circle and graphs of circular functions in this activity.

Materials:

uncooked spaghetti	poster paper	scissors
meter stick	markers	string
glue sticks	protractor	

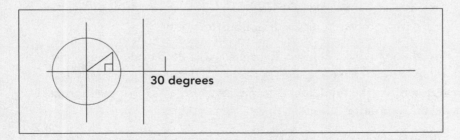

Time to complete: 80 minutes

Please work in groups of 4–6 people

Instructions:

1. At one end of the paper, construct a circle that has a radius of the length of one spaghetti piece, and draw a pair of axes with the origin at the center of the circle.

2. Using a protractor to determine angle measures, mark around the circle every 15 degrees.

3. Adjacent to the circle, draw another set of axes with an x-axis that is about 6.5 spaghetti lengths. Label both axes and also their intersection point as (0, 0). Label the x-axis from 0 to 360° and the y-axis from −1 to 1.

4. Wrap a string around the unit circle. One end of the string should be placed at (1, 0). Use the marker to mark points on the string at each of the special angles (every 15 degrees).

30 degrees

5. The string can now be placed along the horizontal axis with the end of the string that started at the point (1,0) now placed at the origin. The marks that were made on the string will give us the angle measures that we need. Thus the first mark can be transferred to the poster board and labelled 15 degrees.

6. Form a right triangle on the circle in which the radius of the circle to the 15 degree mark is the hypotenuse.

7. Measure a piece of spaghetti that is the same length as the perpendicular line from the circle to the x-axis.

8. Break a piece of spaghetti into an appropriate length for the vertical length of the triangle.

9. Glue the piece of spaghetti onto the paper in order to show the length of the vertical leg of the 15-degree right triangle.

10. Repeat this process for each of the marks around the circle. Think carefully when attempting to build a triangle at 0, 90, 180, and 270 degrees. Can it be done?

11. After all the pieces are glued, draw a smooth curve to connect the dots and answer the questions below. **The graph drawn is one complete sine curve.**

Questions:

1. What is the radius of the circle?

2. In terms of Spaghetti Lengths L, what is the circumference of the circle?

3. Take L = 1 (as spaghetti is the unit here). What is the length of the arc of the circle at 15 degrees? And what is the length at 30 degrees? Fill out the table with the correct values for the arcs:

Angle	15°	30°	45°	60°	75°	90°	180°	270°	360°
Arc Length		$\dfrac{\pi}{6}$							2π

After you have completed the table, plot these values along the horizontal x-axis on your poster. We call these new x-axis units **radians**.

4. Where would a triangle corresponding to 390 degrees (or $\dfrac{13\pi}{6}$ radians) be constructed?

5. What is the **period** of the sine curve? That is, what is the wavelength? After how many degrees or radians does the graph start to repeat?

6. What is the **amplitude** of the sine curve? That is, what is half the difference between the maximum and minimum values on the y-axis?

(Continued)

7. On the poster, separate each quadrant by marking/drawing a rectangle for each 90-degree interval. The first quadrant would have a range from –1 to 1 on the y-axis and a width of domain of one quadrant, which is 90 degrees.

8. Describe whether the sine curve is positive or negative in each quadrant:

Quadrant I .... Quadrant II................................

Quadrant III...................................... Quadrant IV..............................

9. Using your calculator (not spaghetti), find the values of the **cosine** at each degree listed on the x-axis. With a different color marker, mark the points on the paper as you calculate them. After all of the dots have been added, draw a smooth curve to connect them. **The graph drawn is one complete cosine curve**.

10. Describe whether the cosine curve is increasing or decreasing in each quadrant:

Quadrant I .... Quadrant II................................

Quadrant III...................................... Quadrant IV..............................

Extension:

Sine Curve Fitting: $f(x) = a \sin(b(x - c)) + d$

Amplitude (a): The height of the curve or one-half the difference between the maximum value and the minimum value.

Frequency (b): The number of times the function repeats itself in a given interval.

Period $\left(\dfrac{2\pi}{b}\right)$: The interval needed in order to see one full curve, or the length of the interval divided by the frequency. For sine and cosine, the interval is 360°. For tangent, it is 180°.

Horizontal shift (c): The glide of the entire graph to the left or the right.

Vertical shift (d): The glide of the entire graph up or down.

Use your GDC (graphical display calculator or graphing software) to find out the effects of each parameter a, b, c, and d.

FIGURE 7.7: THE UNIT CIRCLE

The Unit Circle

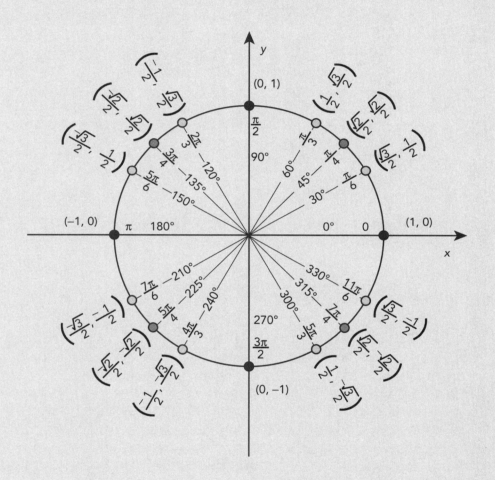

Belk (2010).

This should be copied on a laminate or transparent paper and overlaid on a piece of graph paper.

(Continued)

Unit Circle Graphs: A Paper Calculator

Resources: the unit circle, graph paper, pencil

Sine Curve

Using the unit circle, complete the table below. When you have completed this, plot the points x and y on graph paper and draw a smooth curve through your points.

x	0	15	30	45	60	75	90	105	120	135	150	165	180	195	210	225	240	255	270	285	300	315	345	360
x radians																								
y																								

Cosine Curve

Use the unit circle to help you to complete the table for the cosine curve and plot your x and y points on graph paper.

x	0	15	30	45	60	75	90	105	120	135	150	165	180	195	210	225	240	255	270	285	300	315	345	360
x radians																								
y																								

Tangent Curve

Repeat the above instructions.

x	0	15	30	45	60	75	90	105	120	135	150	165	180	195	210	225	240	255	270	285	300	315	345	360
x radians																								
y																								

The Unit Circle Follow-Up Activity

Look at your three beautiful curves you have plotted with your unit circle and answer the following questions:

11. Describe whether the sine curve is positive or negative in each quadrant:

    Quadrant I .... Quadrant II ....

    Quadrant III .... Quadrant IV ....

From your graph and table you can see sin 30 = 0.5

> What other values of θ can you find for sin θ = 0.5?
>
> What angle is this? Which quadrant is this?
>
> How would I work this out from the graph?
>
> What angle/s give sin θ = −0.5?
>
> How would you work this out from the graph?

12. Describe whether the cosine curve is positive or negative in each quadrant:

    Quadrant I.......................................Quadrant II ....

    Quadrant III.....................................Quadrant IV ....

13. Describe whether the tangent curve is positive or negative in each quadrant:

    Quadrant I.......................................Quadrant II ....

    Quadrant III.....................................Quadrant IV.....................................

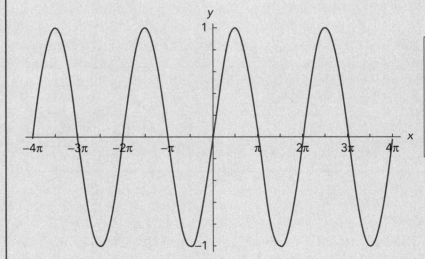

All angles in the first quadrant between 0° and 90° are called principal or base angles.

(Continued)

All Stations to Central

All
Sine
Tangent
Cosine

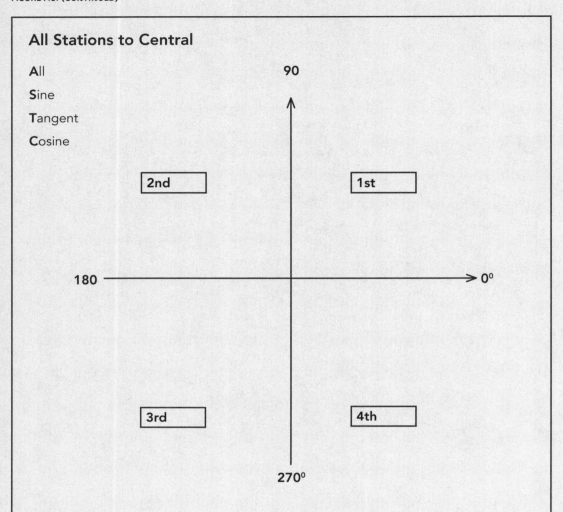

Summarizing your results (you may need your teacher's help here):

Write the first letter of the trigonometric ratio that is positive for each quadrant.

Write down the radian next to each degree.

How would you work out each angle for each quadrant given the base or principal angle?

FIGURE 7.8: TRIANGLE INEQUALITY

Triangle Inequality

Using one piece of spaghetti at a time, break your
spaghetti into three pieces and measure each length.
Do your three pieces make a triangle? Repeat this and
complete the table below:

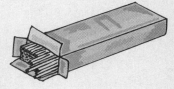

Length A	Length B	Length C	Does this make a triangle?

What do you notice about the relationship of the lengths of spaghetti when a
triangle is formed?

Write a generalization that summarizes the key idea.

Chapter Summary

Developing conceptual understanding in this day and age has the added benefit of our vast technological advances. Technology can assist students to understand concepts without having to manually compute long, tedious calculations.

There are two models that help teachers plan for technology integration: SAMR and TPACK. The SAMR (substitution, augmentation, modification, and redefinition) model provides a scaffold for teachers as they adopt technological tools in their classroom. TPACK provides a framework to allow teachers to consider three knowledge areas—technological, pedagogical, and content—and how they interact with each other when planning for integration of technology. Specialist mathematics software can assist students to form generalizations—statements of conceptual understanding—in an inductive, more efficient and effective fashion. Technology also allows students to see multiple representations using graphing software, which further supports a deeper understanding of the concepts.

The flipped classroom has revolutionized the role of the teacher, moving away from being a "sage on the stage," with class time focused on what students do not understand, to instead being a "guide on the side" who focuses on what they do understand.

Collaborative tools such as Google Docs have revolutionized group work in the classroom where every student's contribution is taken into account and valued. There is a vast array of downloadable applications readily available to support formative assessment and concept mind mapping. These types of tools promote collaboration and allow students to develop their social skills not only in the context of their classroom but on a global level.

The key idea to integrating technology effectively into classrooms is "it's not *what* you use; it's *how* you use it." Ask yourself the question, Does the technological tool enhance learning in your students?

The next chapter will discuss the components of the ideal concept-based math classroom.

Discussion Questions

1. How might you use the SAMR model in your own practice?
2. How might you apply the TPACK model when planning a unit of work?
3. There are so many digital tools on the market, how do you decide which ones to use?
4. How do you integrate technology in the math classroom in order to promote and enhance learning?
5. How do you ensure students still engage in discussions when using Google tools such as Google Docs or Google Slides?
6. When would you use technology in your practice and when would you not?

CHAPTER 8

What Do Ideal Concept-Based Mathematics Classrooms Look Like?

Be the change you want to see in the world.

—Mahatma Gandhi

In one of my first meetings in a new faculty a few years ago, I was asked to share a variety of inquiry-based learning experiences with the goal of presenting different approaches in the classroom. I talked about visible thinking routines and using an inquiry approach to promote deeper conceptual understanding. Someone said, "Our exam results have been very good for the last 20 years. Why do we need to change?" "Here we go," I thought. The old "That's the way we always did it!" TTWWADI is a new acronym that represents this resistance to change! Other commonly heard refrains are, "If it ain't broke, don't fix it," and "What is wrong with what we are doing now?"

I posed the question, "What do ideal math classrooms look like?" All agreed that ideal learning in math looked like this:

Students are highly engaged in the learning and there is excitement in the room.

All students are presented with some challenge, regardless of ability or background.

Students have a deep understanding of the concepts involved and are able to apply their understanding to different contexts.

These features are typical of concept-based curriculum and instruction.

In order to develop concept-based curriculum and instruction further, what would the ideal math classroom look like in the future? The next few sections outline components of the ideal classroom. Common concerns and misconceptions about concept-based curriculum and the use of inquiry will also be addressed.

Foster a Culture of Growth Mindset

The word "change" can elicit negative reactions in people because human beings are naturally resistant to change. I often hear, "Why do I need to change?" A more befitting word that may have less negative connotation would be *evolve*. The process of change should be described as an evolution, implying development and continually seeking improvement of one's craft. Evolution is a natural process that we, as a society, go through to move toward a more complex or better form. This idea is grounded in a growth mindset that with effort we continue to evolve, to increase our intellect, and embrace a love of learning.

> With any introduction of a new framework for teaching and learning, teachers need to be convinced of the value in terms of students' engagement and achievement. Knowing how to address questions that arise will help support the adoption and acceptance of concept-based curriculum.

Fostering a growth mindset (Dweck, 2006) will encourage evolution as this belief is grounded in the principle that, with effort, intelligence and abilities can develop due to the brain's neuroplasticity. This applies to both teachers and students.

In a study of adolescents (Good, Aronson, & Inzlicht, 2003), students who received growth mindset training (compared to matched controls who received other instruction) showed significant improvement, specifically in their math and verbal achievement test scores. It was interesting to note that girls who received the growth mindset training narrowed the gender gap in math achievement.

Growth mindset aligns with the National Council of Teachers of Mathematics (NCTM) *Principles to Actions* (2014): "Support productive struggle in learning mathematics. Effective teaching of mathematics consistently provides students, individually and collectively, with opportunities and supports to engage in productive struggle as they grapple with mathematical ideas and relationships."

An excellent phrase to foster growth mindset with your students is "not yet." Let your students know that they may not have complete understanding of a concept *yet,* but with effort and dedication they will be able to achieve understanding.

Pedagogical Principles in an Ideal Classroom

International systems around the world address the need for conceptual understanding in curriculum and instruction. These include various state standards from the United States as well as standards from Australia, Canada, Hong Kong, and other countries.

The NCTM, internationally regarded for its exemplary math precepts, has released eight Mathematics Teaching Practices. These align very closely with the Standards for Mathematical Practice from the U.S. Common Core State Standards as well as with mathematics teaching approaches from other international curricula.

The eight Mathematics Teaching Practices are outlined below.

Establish mathematics goals to focus learning. Effective teaching of mathematics establishes clear goals for the mathematics that students are learning, situates goals within learning progressions, and uses the goals to guide instructional decisions.

Implement tasks that promote reasoning and problem solving. Effective teaching of mathematics engages students in solving and discussing tasks that promote mathematical reasoning and problem solving and allow multiple entry points and varied solution strategies.

Use and connect mathematical representations. Effective teaching of mathematics engages students in making connections among mathematical representations to deepen understanding of mathematics concepts and procedures and as tools for problem solving.

Facilitate meaningful mathematical discourse. Effective teaching of mathematics facilitates discourse among students to build shared understanding of mathematical ideas by analyzing and comparing student approaches and arguments.

Pose purposeful questions. Effective teaching of mathematics uses purposeful questions to assess and advance students' reasoning and sense making about important mathematical ideas and relationships.

Build procedural fluency from conceptual understanding. Effective teaching of mathematics builds fluency with procedures on a foundation of conceptual understanding so that students, over time, become skillful in using procedures flexibly as they solve contextual and mathematical problems.

Support productive struggle in learning mathematics. Effective teaching of mathematics consistently provides students, individually and collectively, with opportunities and supports to engage in productive struggle as they grapple with mathematical ideas and relationships.

Elicit and use evidence of student thinking. Effective teaching of mathematics uses evidence of student thinking to assess progress toward mathematical understanding and to adjust instruction continually in ways that support and extend learning.

Research shows these eight teaching practices need to be consistent components of every math lesson. Notice the emphasis on conceptual understanding, promoting student thinking, and clear goals to focus learning. The concept-based model aligns well with these practices and represents a thinking model in which generalizations are the focus for learning.

The International Baccalaureate (2015) has published its Approaches to Teaching and Learning. The six pedagogical principles underpinning the IB diploma program are listed below.

Teaching in the IB program is

1. based in inquiry;
2. focused on conceptual understanding;
3. developed in local and global contexts;
4. focused on effective teamwork and collaboration;
5. differentiated to meet the needs of all learners;
6. informed by assessment.

These principles are consistent with the tenets of concept-based curriculum and instruction as world education systems recognize the need to foster deep conceptual understanding in our students to prepare them for future success. Several of these principles are covered, with examples, in Chapter 5.

Developing the Ideal Concept-Based Mathematics Lesson

Lois Lanning (Erickson & Lanning, 2014) has developed a rubric to help teachers identify stages of the developing concept-based teacher in terms of concept-based instruction. The ideal math lesson would see the teacher at mastery level for the lesson opening, targets during the lesson, and the lesson closing. These are described in detail below.

The Lesson Opening

✓ The opening clearly communicates an engaging, captivating overview of the lesson that connects and extends previous learning.
✓ The lesson immediately engages students' minds and interests.

I call this the "hook" of the lesson where you engage student interest, curiosity, and motivation from the very beginning. Rob Colaiacovo, Prince Alfred College, Australia is an experienced lead teacher who utilizes the inquiry approach to help students understand the concepts in mathematics. Here is one of his activities, which is an example of using a hook and a focus on conceptual understanding.

He poses a question that seems to be so simple, yet provokes thought and curiosity. Research from Gelman et al. (2014) shows that learning is improved when students are curious.

I really like to use this activity to help students to understand the concept of "for every" when learning about fractions. The following question is posed:

Cathy gives away 20 chocolate frogs and Peter gives away 30 chocolate frogs. Who is more generous?

Pretty quickly you get students in one of three categories:

1. Peter because he gives away more (approx. 70%).
2. Not sure because we don't know how many each has (25%).
3. Cathy because it's a trick question (5%).

We establish the idea that it's not only how many you give away that's important, but also how many you may actually have.
We then move on to some simple activities until we reach the following problem:

Cathy has 90 chocolate frogs and gives away 80 whilst Peter has 40 chocolate frogs and gives away 30. Who is more generous?

You see, this whole unit is to get students to think about fractions in the "for every" sense rather than a part /whole approach—even though the two can be linked.

For example, if Peter has 49 frogs and gives away 14, he is giving away 2 for every 7 that he has.

Can you see that in the diagram below?

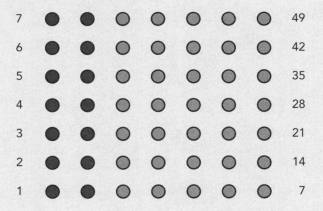

(Continued)

(Continued)

Anyway, students who do not think about this problem multiplicatively will say that both people are equally generous because they both have 10 left.

When thinking about fractions in the "for every" sense, we need to know many frogs Cathy or Peter starts with. If Cathy has 90 frogs and gives away 80 frogs, then she is giving away 8 for every 9 she has. Whereas if Peter has 40 frogs and gives away 30, he is giving away 3 for every 4.

One student, David (not his real name), in particular, was absolutely adamant that they are both equally generous and could not understand how Cathy was more generous until I showed him (and explained) this picture.

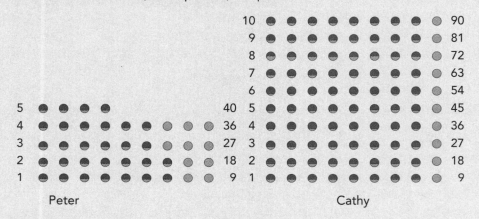

I tell him can he see from the diagram that Cathy is giving away 8 for every 9 she has? A light bulb moment comes over David and he says, "Yes!"

Now, can you see that Peter is giving LESS than 8 for every 9 he has?

David looks at it for a very long time and says to me, "So just because they end up with the same amount doesn't necessarily mean they're equally generous?" I smiled at him.

It was a great lesson.

We do all the formal stuff with fractions later!

Targets During the Lesson

✓ The execution of the lesson plan is well paced, while being flexible and responsive to anticipated student needs (and needs that arise throughout the lesson).

✓ There is conscious and consistent development of students' synergistic thinking through instructional techniques and thought-provoking examples and through learning experiences and resources that bridge to a deeper, conceptual idea (understanding).

- ✓ Teaching uses a variety of techniques to support the transfer of learning and deepening of understanding (e.g., questioning, asking for other examples/non-examples of the same concept or generalizations, feedback, and asking students for an analysis of their reasoning with supporting evidence).
- ✓ Gradual release of responsibility and ownership for learning from teacher to student is clear.
- ✓ There is continuous monitoring of students' independent and collaborative group work with timely, relevant feedback and questions that facilitate and mediate the learning process.

Lesson Closing

- ✓ Evidence of learning (formative or summative) the lesson's targeted knowledge, skills, and understandings is collected.
- ✓ Collaboratively the teacher and students reflect and analyze the success of the learning (process and product).
- ✓ Students learn how the learning will build toward future learning targets.

Figure 8.1 outlines Lois Lanning's (Erickson & Lanning, 2014) full rubric for developing concept-based instruction. Use the rubric to assess your own level of mastery. Once you determine where you are in the rubric, devise a plan for improving your skills to the next level.

Developing Concept-Based Lesson Planning in the Ideal Mathematics Classroom

Lois Lanning (Erickson & Lanning, 2014) has also created rubrics to help develop concept-based lesson planning. The main components of a concept-based lesson plan at mastery level are described in Figure 8.2. Figure 8.2 outlines our ultimate goals for concept-based lesson planning and important points to keep in mind when planning an ideal lesson. Knowing our end point, our goals align with the philosophy of Wiggins's and McTighe (2006a, 2006b) work on Understanding by Design.

Figure 8.3 shows the entire rubric for concept-based lesson planning. This may be used when embarking on your journey toward concept-based lesson planning. Think about each component of the lesson plan and identify where you are on this rubric. Set targets to see how you can progress from novice to emerging to mastery level.

Download and complete Figures M8.4–M8.8 on the companion website to assess your level (novice, expert, or mastery) in developing concept-based instruction, writing concept-based lesson plans, and understanding concept-based curriculum and instruction. Use these templates to also set targets for yourself to get to the next level.

FIGURE 8.1: RUBRIC: CONCEPT-BASED INSTRUCTION BY DR. LOIS A. LANNING

	Novice	Emerging	Master
Lesson Opening	• Lesson opens by launching directly into an activity without providing an overview or clear directions • The lesson's target generalization is posted or stated at the beginning of the lesson rather than drawn from students throughout • The lesson opening is accurate but bland and leaves students disinterested as it is more of a teacher's monologue	• The lesson opening sets the stage for inductive teaching that will draw out conceptual understandings from the students (e.g., examples are posted, intriguing questions presented, an interesting scenario shared, relevant concepts posed) but the opening is overly detailed and too long	• The opening clearly communicates an engaging, captivating overview of the lesson that connects and extends previous learning
During Instruction reflects modeling, facilitating, and mediating conceptual understanding	• The lesson follows the written lesson plan but since the plan does not address all the elements of an effective concept-based lesson, instruction falls short • Concepts are somewhat apparent in the lesson with little attention to how to use concepts to create intellectual engagement and deepen students' understanding • Focus is less on the transfer of learning and more on task completion	• Follows the concept-based lesson plan prescriptively • Maintain the emotional engagement of students by using examples and resources • There is some gradual releases of responsibility of learning from teacher to student, but the teacher is assuming most of the cognitive work	• The execution of the lesson plan is well paced, while being flexible and responsive to anticipated student needs (and needs that arise throughout the lesson) • There is a conscious and consistent development of students' synergistic thinking through instructional techniques and thought provoking examples and through learning experiences and resources that bridge t a deeper, conceptual idea (understanding)

222

	Novice	Emerging	Master
	• Instruction employs different kinds of questions as the main tool for encouraging the transfer of learning but still over-relies on factual questions • Instruction remains predominantly teacher centered • Student participation is predominantly in response to teacher questioning and evaluation	• Most student's are engaged in the learning while clusters of students may remain off task or disinterested due to ineffective level of challenge or lack of relevance	• Teaching uses a variety of techniques to support the transfer of learning and deepening of understanding (e.g. questions, asking for other examples/non-examples of the same concept or generalization, feedback, and by asking students for an analysis of their reasoning with supporting evidence) • Gradual release of responsibility and ownership for learning from teacher to student is clear • Continuous monitoring of students' independent and collaborative group work with timely relevant feedback and questions that facilitate and mediate the learning process
Lesson Closing	• The teacher recaps the learning experiences in the lesson	• There is closing assessment (formative or summative) of the knowledge and skills students learned and an attempt to determine students' level of conceptual understanding • Relevant practice beyond the lesson is assigned	• Evidence of learning (formative or summative) the lesson's targeted knowledge, skills, and understanding is collected • Collaboratively the teacher and students reflect on and analyze the success of the learning (process and product) • Students learn how the learning will build future learning targets

Transitioning to Concept-Based Curriculum and Instruction, Corwin Press Publishers, Thousand Oaks, CA

Lesson opening	The lesson opening engages synergistic thinking by asking students to consider the knowledge and/or skills they will be learning through a conceptual question or lens.
Learning targets	Learning targets represent what students are expected to know, understand, and do; the limited number of learning targets allows for in-depth, focused instruction and learning.
Guiding questions	Potential questions are of different types (factual, conceptual, provocative) and are listed throughout the plan. The lesson plan shows a deliberate effort to use questions to help students bridge from the factual to the conceptual level of understanding.
Learning experiences	The student work requires students to cognitively wrestle with and synthesize the knowledge, skills, and concepts under study in relevant contexts that lead to the realization of the generalization. Student work is at the appropriate level of challenge; is intellectually and emotionally engaging, meaningful, and relevant to the discipline; and provides appropriate student choice. The learning experiences are deliberately designed to enhance the transfer of learning across other disciplines and situations.
Assessment methods	Assessment types are varied so they assess students' developing knowledge, skills, and understandings (generalization) and allow for timely feedback. Assessments provide relevant information about students' process of learning as well as their learning products. Student self-assessment is valued.
Differentiation	Plans for differentiation to meet the needs of all learners are included and support all students' meeting a common conceptual understanding (generalization). Differentiation is based on an analysis of multiple data points that reveal individual student learning needs. Specific accommodations are readily available based on anticipated student misconceptions and needs.
Lesson design	The lesson design is primarily inductive, requiring students to engage in a multifaceted inquiry process and to reflect on the connections across the examples presented so that students can formulate and defend their generalizations. A deductive design may also be included to support the learning of foundational facts and skills.
Closing	

© 2014 Lois A. Lanning

Transitioning to Concept-Based Curriculum and Instruction, Corwin Press Publishers, Thousand Oaks, CA

FIGURE 8.3: THE DEVELOPING CONCEPT-BASED TEACHER: CONCEPT-BASED LESSON PLANNING

Components of Lesson Plan	Novice	Emerging	Master
Lesson opening: An explicit and engaging summary of the work to be accomplished that triggers synergistic thinking	Lesson opens by stating the activities students will experience in the lesson and may include stating the generalization the lesson will be teaching toward	Lesson opening contains a conceptual lens but the weak tie to content fails to engage synergistic thinking.	The lesson opening engages synergistic thinking by asking students to consider the knowledge and/or skills they will be learning through a conceptual question(s) or lens
Learning targets: What students are expected to know (factual knowledge), understand (generalization), and be able to do (skills)	What students must Know and/or be able to Do) is listed in the lesson plan	Targets identify what students must Know, Understand (generalization), and Do, but there may be more learning targets than can be accomplished in-depth within the lesson timeframe	Learning targets represent what students are expected to Know, Understand, and Do; the limited number of learning targets allows for in-depth, focused instruction and learning
Guiding questions: The three different types of questions (factual, conceptual, provocative) serve as a bridging tool for conceptual thinking and problem solving.	Questions in the lesson plan focus heavily on factual knowledge and routine skills	Lesson questions reflect different types (factual, conceptual, and possibly provocative questions) and anticipate student misconceptions	Potential questions are of different types (factual, conceptual, provocative) and are listed throughout the plan
Learning experiences: Intellectually engaging student work that provides opportunities for students to practice their learning and to arrive at the target generalization (conceptual understanding)		The student work attempts to pursue conceptual understandings but may not provide a clear pathway with enough examples or scaffolds for students to realize the conceptual understanding (generalization) Lesson plan shows efforts to engage student work to engage students' interest and offers some student choice	The lesson plan shows a deliberate effort to use questions to help students bridge from the factual to the conceptual level of understanding The student work requires students to cognitively wrestle with and synthesize the knowledge, skills, and concepts under study in relevant contexts that lead to the realization of the generalization

(Continued)

Components of Lesson Plan	Novice	Emerging	Master
Assessment methods: Assessment types are selected according to the lesson's learning targets (know, understand, do) and to the assessment purposes (formative & summative) in order to capture evidence of students' learning (process & product) which will then inform instruction.	The targeted knowledge and skills are indicated in the plan, but the learning experiences do not require students to apply their learning in relevant contexts, which would clearly lead to conceptual understanding and transfer across learning situations	Assignments are intellectually and emotionally engaging but are not at the appropriate level of challenge for all students Assessment types are varied and help monitor students developing knowledge and skills Assessments of understanding, aligned to the target generalization, are not clear	Student work is at the appropriate level of challenge, is intellectually and emotionally engaging, meaningful, and relevant to the discipline, and provides appropriate student choice The learning experiences are deliberately designed to enhance the transfer of learning across other disciplines and situations Assessment types are varied so they assess students' developing knowledge, skills, and understandings (generalization) and allow for timely feedback Assessments provide relevant information about students' process of learning as well as their learning products Student self-assessment is valued Plans for differentiation to meet the needs of all learners are included and support all students' meeting a common conceptual understanding (generalization)
Differentiation: Lesson adjustments are planned, as needed, for the content students are expected to master, the process students will use to access the content, and the product students will produce to show their learning. The conceptual understanding (generalization) all students are expected to realize remains consistent for all students.	The student work required in the lesson primarily relies on worksheets, disconnected skills and facts that are not authentic or intellectually/emotionally engaging for students		

Components of Lesson Plan	Novice	Emerging	Master
Lesson design: In a deductive lesson design, the teacher states the learning targets (including generalization) to the learners at the beginning of the instruction. In an inductive design, students construct their understandings through an inquiry process. Closing: Plan for a way that evidence of learning can be reviewed collectively	Assessment types are limited so it is difficult to know the degree of students' learning and progress toward conceptual understanding Plans for differentiation may be stated but lack relevance to individual student learning needs The lesson design is deductive (e.g., objective to example vs. example to generalization)	Plans for differentiation are included for students who need support (e.g., special education, ELL) in the areas of content, process, and product Misconceptions are generally addressed with the class as a whole The lesson design attempts to use inductive teaching but the examples presented only vaguely illustrate the targeted conceptual understandings A deductive design may also be included to support the learning of foundational facts and skills	Differentiation is based on an analysis of multiple data points that reveal individual student learning needs Specific accommodations are readily available based on anticipated student misconceptions and needs The lesson design is primarily inductive, requiring students to engage in a multifaceted inquiry process and to reflect on the connections across the examples presented so that students can formulate and defend their generalizations A deductive design may also be included to support the learning of foundational facts and skills

Transitioning to Concept-Based Curriculum and Instruction, Corwin Press Publishers, Thousand Oaks, CA.

Common Concerns and Misconceptions About Concept-Based Curriculum and Instruction

The journey to concept-based curriculum and instruction is one that will not be without challenges. Importantly, adopting a growth mindset will make this journey smoother and allow room for the belief that with effort everyone can do math when instruction is focused on conceptual understanding.

As you embark on your concept-based journey, you may ask questions or come across common concerns and misconceptions about concept-based curriculum and instruction, which are discussed below.

I cannot adopt concept-based curriculum and instruction because I have to follow a prescribed curriculum and I am not a curriculum designer!

Concept-based curriculum is an approach and a philosophy borne of the idea that conceptual understanding of math is vital to developing students' intellects. It does not necessarily dictate instruction and can overlay any existing curriculum. It's a way of organizing information so that students ultimately have a deeper conceptual understanding of the mathematical content. This aligns with the NCTM *Principles to Actions* (2014): "Build procedural fluency from conceptual understanding. Effective teaching of mathematics builds fluency with procedures on a foundation of conceptual understanding so that students, over time, become skillful in using procedures flexibly as they solve contextual and mathematical problems."

> The concept-based approach does not necessarily dictate instruction and can overlay any existing curriculum.

I cannot adopt concept-based curriculum and instruction because it doesn't prepare students for college and the real world.

With the exponential growth of information and the digital revolution, the traits to be successful in this modern day and age require efficient processing of new information and a higher level of abstraction. Concept-based curriculum and instruction fosters key skills to prepare students for future success. These key skills include critical thinking, problem solving, reasoning, and the ability to transfer and apply concepts to different contexts.

This approach will prepare students with the critical thinking skills necessary for college as well as engage students intellectually and emotionally. Concept-based curriculum and instruction provides rigor by challenging student thinking.

Hart Research Associates (2013) report that the following are the top skills that employers seek:

- Critical thinking and problem-solving ability;

- Collaboration—the ability to work in a team;

- Communication—oral and written;

- The ability to adapt to a changing environment.

> **This approach will prepare students with the critical thinking skills necessary for college as well as engage students intellectually and emotionally.**

I cannot adopt concept-based curriculum and instruction because I have to teach to an exam!

In reality, testing is a part of many education systems, so it is not something one can avoid. How do you prepare your students for final examinations? In order to be able to answer questions on an exam, students need to have a deep, thorough understanding of the material, that is, a deep con-

> **Research has shown that concept-based instruction leads to higher scores than traditional environments.**

ceptual understanding that they can apply to any problem posed. Research has shown that concept-based instruction leads to higher scores than traditional environments (Gao & Bao, 2012). Concept-based curriculum and instruction develops students' intellect by stimulating synergistic thinking: higher order thinking that connects the factual level and the conceptual levels.

The Story of Mei

A few years ago, I had a very conscientious, able student join my high school class. Mei (not her real name) had experienced years of traditional math teaching with great success. Mei said she was completely lost for the first few math lessons with me. She asked me, "Can you just show me the formula, give me some examples, and then I will work on exercises from my textbook." This was the approach she was used to.

In my math lessons I challenged her to think; I did not provide answers but questions to support her understanding. I counseled Mei and told her to give herself time to adjust. After three months, she scored an A+ for her in-class assessment, and she said she hardly invested any time to study for the test. She said all the concepts were just there in her brain, and she was able to apply her understanding to any question. In the past, she had to commit a lot of time to reviewing. Mei said that math lessons were now fun, and she understood why my math lessons focused on tasks that helped

(Continued)

with her conceptual understanding rather than rote memorization of algorithms or completing dozens of questions.

Adults find change difficult, as do our students. Students need time to adjust to a different approach, but if your activities are motivating, engaging, and give students opportunities to communicate their understanding, this transition will be smoother.

I don't have time for inquiry! I need to get through the content.

Utilizing inquiry is a reallocation of time, which simultaneously covers the content. It involves adopting a different mindset about what productivity looks like in the math classroom. By transitioning from assigning a set of math problems for students to complete to, instead, assigning students one quality performance task to work on, teachers are able to promote a deeper understanding of the math concept. This approach is different from simply teaching an algorithm to be memorized and repeated without ensuring a conceptual understanding of the content. The goal here is to facilitate deep, transferable, and conceptual understanding of a topic. Once this deep understanding of the concept is achieved, it is not necessary for students to complete dozens of questions or problems. Instead, they can spend valuable class time applying their new understanding to multiple, varied contexts, thus demonstrating and deepening their understanding. When the practice of process skills is required, this can be done outside of the classroom. Content is not sacrificed as this is covered simultaneously, with a focus on conceptual understanding. Prince and Felder (2007) stated that problem-based learning through inquiry is a powerful strategy that promotes active learning and a student-centered approach.

I have inquiry lessons once per week.

There is a misconception that inquiry should be a standalone activity used once per week. Inquiry is a necessary component of every lesson and the reason why the IB has included inquiry as one of the underpinning pedagogical principles.

The goal of math lessons is to promote the development of critical thinking, problem solving, reasoning skills, and other higher order skills. The development of these key skills is important in every math lesson. Inquiry is the vehicle that allows us to develop these important critical thinking skills and, therefore, develop intellect as a result.

Inquiry just doesn't work with my students, as they need to be spoon fed! My students are different; they need to be told the answers.

Casner-Lotto and Benner (2006) found the top skills employers look for in their employees are the following:

Critical thinking and the ability to problem solve;

Information technology application;

Teamwork/collaboration;

Creativity/innovation.

These skills will develop in an environment in which students are given the opportunity to be creative and independent. Instead of always providing answers, provide good questions that allow students to think for themselves. The art of good questioning comes from the teacher's ability to draw out understandings and guide students to the big, essential ideas of a topic. Brigid Barron and Linda Darling-Hammond (2008) found that, "Research shows that such inquiry-based teaching is not so much about seeking the right answer but about developing inquiring minds, and it can yield significant benefits. For example, in the 1995 School Restructuring Study, conducted at the Center on Organization and Restructuring of Schools by Fred Newmann and colleagues at the University of Wisconsin, 2,128 students in twenty-three schools were found to have significantly higher achievement on challenging tasks when they were taught with inquiry-based teaching, showing that involvement leads to understanding. These practices were found to have a more significant impact on student performance than any other variable, including student background and prior achievement."

> Spoon-feeding your students may not benefit them in the long term and may inhibit your students from developing into independent, lifelong learners.

Utilizing the levels of inquiry allows students to access the material with different levels of support. Inquiry is about asking quality questions that make students think critically. If students are used to being spoon fed answers and have been exposed to a traditional approach to math learning, then a transition must be carefully planned. Good questioning aligns with the NCTM's *Principles to Action* (2014): "Pose purposeful questions. Effective teaching of mathematics uses purposeful questions to assess and advance students' reasoning and sense making about important mathematical ideas and relationships."

Inquiry does not work for my students, as they all have varying abilities.

A growth mindset believes that with effort, every student is capable of doing math. The levels of inquiry exist for different students with different backgrounds and entry levels. Differentiation is key to being able to provide opportunities for every student to learn in your classroom. Knowing your students and when they need more scaffolding at different times will ensure every student is making progress. In the words of Carol Ann Tomlinson and Susan Demirsky Allan (2000), "Differentiation is more than a strategy or series of strategy . . . it is a way of thinking about teaching and learning" (p. 13).

Inquiry develops critical thinking skills and the ability to inquire and investigate on an independent level. These are skills that we want to nurture in all of our students regardless of their background. This aligns with the NCTM *Principles to Actions* (2014): "Elicit and use evidence of student thinking. Effective teaching of mathematics uses evidence of student thinking to assess progress toward mathematical understanding and to adjust instruction continually in ways that support and extend learning."

> **Differentiation is key to being able to provide opportunities for every student to learn in your classroom.**

Fostering conceptual understanding through the synergistic relationship between the facts, skills, and concepts will encourage development of intellect in our students. Ultimately, we hope our students will be able to apply and transfer their knowledge and skills to prepare them to be global citizens of the 21st century.

Concept-based curriculum makes sense and I like how it is a thinking curriculum and model that could be used with any standards. I particularly loved the inquiry aspect and how students are encouraged to think conceptually, be empowered to be in control of their own learning and as a result develop into independent, lifelong learners with the ability to apply higher order thinking skills across situations.

Ann Marie Cullinan, Chief Academic Officer
Waterbury Public Schools, Connecticut

Chapter Summary

What do ideal math classrooms look like? An important feature is fostering a growth mindset to encourage your students to not be afraid to take risks and recognize mistakes are part of the learning process. A growth mindset believes that everyone is capable of doing and understanding math. Math is not for the elite few.

Education systems around the world recognize the importance of developing conceptual understanding. Pedagogical principles outlined by the NCTM's *Principles to Actions* (2014) and the International Baccalaureate's *Approaches to Teaching and Learning* (2014) include developing conceptual understanding as a key component of mathematics lessons.

Lois Lanning (Erickson & Lanning, 2014) has designed rubrics to support the development of a concept-based teacher. One rubric shows the development of concept-based instruction and includes the mastery levels for a lesson opening, lesson duration, and lesson closing. Another rubric shows mastering developing concept-based lesson plans and includes the components lesson opening, learning targets, guiding questions, learning experiences (in terms of assessment methods, differentiation, and lesson design), and lesson closing. These rubrics can be used to monitor individual and staff journeys toward implementing a concept-based curriculum and instructional model.

Common concerns and misconceptions about concept-based curriculum and instruction were addressed. In summary, when you are able to implement concept-based teaching at the mastery level, you will find that your reasons for resisting the implementation of inquiry-based learning will have subsided once you see the significant learning benefits for your students. You will also find that your concerns will no longer be an issue because the concept-based instructional model will result in deeper, longer-lasting learning.

Last Words

Math education needs to undergo a revolution. Traditional didactic approaches promote memorization of facts and skills with little attention paid to the understanding of mathematical concepts and their relationships. Many jobs will no longer exist in the next two decades, so how will our students be able to stand out? How will you prepare your students for the future? Education systems around the world

(Continued)

(Continued)

need to rethink how to prepare students for an unknown future. We have no idea what technological advances will be invented in the 21st century; however, we do know that transferable attributes such as critical thinking, creativity, and conceptual understanding will be of great importance. The research citing the benefits of concept-based curriculum in mathematics is overwhelming. Conceptual understanding develops higher-order thinking and transferable skills, and it prepares our students for an exciting not-too-distant future.

In summary, the five critical aspects of concept-based pedagogy extended from the work of Erickson and Lanning (2014) are as follows:

1. Synergistic thinking: This is at the heart of concept-based instruction and requires teachers to foster the synergistic relationship between the factual and conceptual levels of thinking.

2. Three-dimensional curriculum design: This includes generalizations and principles—statements of conceptual understanding and the goals for our students after their program of study.

3. The conceptual lens: This helps focus a unit of work and ensures synergistic thinking.

4. Inductive teaching: This develops students' ability to generalize from specific examples.

5. Guiding questions: These are the factual, conceptual, and debatable questions for a unit of work. Through inquiry, these questions draw out essential understandings from a unit of work.

In this ever changing, dynamic, and complex world, mathematics education must engage students intellectually and emotionally. The ability to think conceptually, transfer understandings across contexts and situations, and to enjoy learning and problem solving are major goals for mathematics education today so we can prepare our students for future success.

I hope this book inspires you on your journey to develop conceptual understanding in your students, to eradicate math anxiety and fear by fostering a growth mindset, and to join me in this much-needed math education reform. I hope that eventually every student sees the beauty in math as a result of positive learning experiences in classrooms that are focused on conceptual understandings and equipping students with the necessary skills for an unknown, exciting future as a global citizen.

Discussion Questions

1. How will you foster a growth mindset in your classroom?

2. How will you use the rubric for developing as a concept-based teacher and developing concept-based instruction in your own practice?

3. What does ideal math learning look like in your classroom?

4. What does ideal math teaching look like in your classroom?

5. What is the role of a teacher? Is it to transmit information and knowledge? Or is it to inspire, guide, and facilitate the social process of learning?

6. How will you lead the transition toward concept-based math curriculum for teaching deeper understanding?

7. How will you prepare your students for an unknown future?

Glossary

Algorithm: A set of computational rules to produce a specified outcome. PEMDAS (parentheses, exponent, multiplication, division, addition, and subtraction) is an example of an algorithm. (Chapter 2)

Communicating: One of the six important mathematical processes that can be represented on the Structure of Process. The process of communicating in mathematics refers to talking and writing about mathematics. This includes all forms of verbal and nonverbal communication, such as number talks, sharing methodologies through group discussions, and writing mathematics in the form of words and explanations. (Chapter 2)

Concept-based curriculum: Three-dimensional design models that focus on what students will know (factually), understand (conceptually), and be able to do (skillfully). (Chapter 1)

Concepts: Mental constructs or organizing ideas that are timeless, universal, and transferable across time or situations. Mathematics is an inherently conceptual language, so all concepts in math are timeless, universal, and transfer across time or situations. Examples include functions, trigonometry, and statistics. (Chapters 1 and 2)

Conceptual lens: A vehicle that sets up a synergy between the factual and the conceptual processing centers of the brain. The conceptual lens is a broad, integrating concept that focuses a unit of work to allow students to process the factual information. Examples of conceptual lenses include change (this could be used when we study differentiation in calculus), accumulation, structure, form, and order. (Chapters 2 and 4)

Creating representations: One of the six important mathematical processes that can be represented on the Structure of Process. Representations include using schematic representations, graphs, or tables. Creating representations helps learners to understand in different ways and see connections. (Chapter 2) See also **Schematic representations; Pictorial representations**.

Deductive approach: A teacher-centered approach in which students are provided with the conceptual understandings at the beginning of the learning cycle. (Chapter 1) See also **Inductive approach**.

Facts: Specific examples of people, places, situations, or things. They are locked in time, place, or situation. Facts are not transferable and include definitions, memorized formulae in the form of symbols (e.g., $y = mx + c$), or memorizing the different names of polygons (pentagon, hexagon etc.). (Chapters 1 and 2)

Formula: An equation that uses mathematical symbols or variables to show a relationship and is represented by the facts in the Structure of Knowledge. It is a mathematical relationship or rule expressed in symbols. This is a type of shorthand for mathematicians to use when solving problems or to summarize an idea. (Chapter 2)

Generalizations: Two or more concepts stated in a sentence of relationship. They are understandings that transfer through time, across cultures, and across situations. (Chapters 1, 2, and 3)

Growth mindset: A belief that people's most basic abilities can be developed through dedication and hard work—brains and talent are just the starting point. This well-researched belief was developed by psychologist Carol Dweck and popularized in her 2007 book, *Mindset*. (Chapter 5)

Guided inquiry: Encouraging students to utilize different lines of investigation with a predictable outcome. An example of this would be to ask students to explore different proofs for the Pythagorean theorem. Students explore different lines of investigating proofs, but the outcome is the same. (Chapter 1)

Inductive approach: An approach to instruction that opposes deductive teaching. The inductive approach encourages learners to start with specific numerical examples to form generalizations. The difference between the two approaches can be summarized by the role of the teacher and the learning process. Inductive teaching is a student-centered approach in which teachers present learners with specific examples at the beginning of the learning cycle, and teachers guide students to conceptual understandings. (Chapter 1)

Investigating: One of the six important mathematical processes that can be represented on the Structure of Process. This process includes inquiry approaches and working with unfamiliar situations and exploring different lines of inquiry. Other examples of investigating could include researching and being able to extract relevant information. (Chapter 2)

Macro concepts: Our broader mental constructs that encompass a huge body of topics, such as geometry, algebra, statistics, and probability. Macro concepts give us breadth of understanding. (Chapter 2)

Making connections: One of the six important mathematical processes that can be represented on the Structure of Process. This process refers to learners' ability to see connections between facts and how they relate to one another; make connections between symbols and procedures; make connections between what they are learning and the real world; and connect new problems to old. (Chapter 2)

Mathematical processes: Complex, sophisticated performances that consist of strategies, algorithms, and skills. Mathematical processes are broad techniques that students draw upon when learning mathematics and support the understanding of the concepts in a unit of work. Examples of mathematical processes include investigating, making representations, and making connections. (Chapter 2)

Meso concepts: Mental constructs that consist of smaller collections of topics, such as differentiation, integration, functions, or bivariate analysis. *Meso* comes from the Greek word *mésos*, which means "middle." Meso concepts are types of concepts in between micro and macro concepts. For example, the macro concept "calculus" contains two meso concepts: differentiation and integration. (Chapter 2)

Micro concepts: Mental constructs that are smaller than meso concepts. Examples of micro concepts include slope, quadratic, and variable. Micro concepts give us the depth of understanding in mathematics as they provide the detail in a unit of work. (Chapter 2)

Number talks: A communication tool to help students develop computational fluency by comparing number relationship and looking at the different ways to add, subtract, multiply, and divide. Number talks are initiated by oral conversations. Written and graphical representations of mathematical solutions can help support a number talk. (Chapter 2)

Open inquiry: Encouraging students to utilize different lines of investigation, resulting in an unpredictable outcome. (Chapter 1)

Overarching generalizations: Understandings that are beyond the specifics of a unit (e.g., "Mathematics reveals patterns that might have remained unseen"). (Chapter 2)

Pictorial representations: Drawings that do not help solve the problem and are irrelevant to the problem posed. (Chapter 2) See also **Creating representations**.

Principles: Conceptual understandings with the same criteria as generalizations, but they rise to the level of law, theorem, or foundational truth based on the best evidence to date. Principle statements never contain a qualifier such as *often, can,* or *may*. In mathematics, principles are more specific to the unit of study. The Pythagorean theorem is an example of a principle. (Chapters 1, 2, and 3) See also **Generalizations**.

Problem solving: One of the six important mathematical processes that can be represented on the Structure of Process. The fundamental building block of mathematics is problem solving. It is a defined series of skills and strategies that form mathematical techniques that a learner employs when faced with an unfamiliar situation. (Chapter 2)

Processes: Actions that produce results. A process is continuous and moves through stages during which inputs (materials, information, people's advice, time, etc.) may transform or change the way a process flows. A process defines what is to be done—for example, the writing process, the reading process, the digestive process, the respiratory process, and so on. (Chapter 1)

Reasoning and proof: One of the six important mathematical processes that can be represented on the Structure of Process. This process refers to the ability to make generalizations and provide explanations and justifications for arguments. (Chapter 2)

SAMR: An acronym for the hierarchical levels of technology use: substitution, augmentation, modification, and redefinition. (Chapter 7)

Schematic representations: Purposeful visual diagrams, codes, or symbols that help solve a problem. (Chapter 2) See also **Creating representations**.

Skills: Small operations or actions that are embedded in strategies and, when appropriately applied, "allow" the strategies to work. Skills underpin strategies. Examples of skills are being able to create a table of values, plot a graph, or to use trial and error to solve a problem. (Chapter 2)

Social constructivism: A form of cognitive constructivism that emphasizes the collaborative nature for effective learning. Social constructivism was developed by psychologist Lev Vygotsky. This philosophy believes learning is a product of social interactions, with learners integrated into a knowledge community for optimal learning. This book advocates that when students talk to each other about math (thereby socially constructing their understanding of math) in an effort to come to deeper understandings, their learning improves. (Chapter 2)

Strategies: A number of skills that learners use in a methodical and systematic way to support learning. (Chapters 1 and 2)

Structure of Knowledge: A graphical representation of the relationship between the topics and facts, the concepts that are drawn from the content under study, and the generalization and principles that express conceptual relationships (transferable understandings) that are supported by the facts. (Chapters 1 and 2)

Structure of Process: The complement to the Structure of Knowledge, it is a graphical representation of the relationship between the processes, strategies, and skills and concepts, generalizations, and principles in process-driven disciplines like English language arts, the visual and performing arts, and world languages. (Chapters 1 and 2)

Structured inquiry: Heavily scaffolded prompts guiding students to a particular line of investigation with a predictable outcome. An example of this would be to guide students through one particular proof of the Pythagorean theorem. (Chapter 1)

Synergistic thinking: Occurs when the factual level of thinking interacts with the conceptual level of thinking in the three-dimensional concept-based design model, resulting in a deeper understanding and the ability to transfer ideas to other times, places, or situations. The following generalization would be an example of developing synergistic thinking: "Utilizing algebraic tools such as algebraic multiplication, subtraction, and division allows highly complex problems to be solved and displayed." (Preface, Chapters 1, 4, 6, and 8)

Teacher talk time (TTT): The time during which a teacher talks to the whole class. (Chapter 5)

Theorems: Statements in mathematics that have been proven on the basis of previously established statements (theorems). In the Structure of Knowledge, theorems are categorized under principles. The Pythagorean theorem is an example of a principle. (Chapter 2)

Theory: A system of conceptual ideas that explain a practice or phenomenon. Examples include the Big Bang theory and Darwin's theory of evolution. (Chapter 1)

Three-dimensional model of instruction: An educational model that suggests a more sophisticated design with a third level: the conceptual level. In a three-dimensional curriculum and instruction model, the lower levels of the Structure of Knowledge and the Structure of Process (facts, skills, and strategies) are important components, but the third dimension of concepts, principles, and generalizations ensures that conceptual thinking and understanding are prominent. A three-dimensional, inductive approach encourages students to construct generalizations at the end of the learning cycle through the use of inquiry. (Chapters 1 and 2)

Topical generalizations: Subject-specific generalizations (e.g., "Statistical analysis and graphic displays often reveal patterns in seemingly random data or populations, enabling predictions"). (Chapter 2)

Topics: In the Structure of Knowledge, topics organize a set of facts related to specific people, places, situations, or things. Unlike history, for example, mathematics is an inherently conceptual language, so "topics" in the Structure of Knowledge are actually broader concepts, which break down into micro-concepts at the next level. (Chapters 1 and 2)

TPACK: A popular framework that explains how teachers' understanding of educational technologies and their pedagogical and content knowledge interact with one another to produce effective teaching with technology. All teachers have in their toolbox content knowledge and pedagogical knowledge. This is known as CPK (content pedagogical knowledge, from the work of Lee Shulman, 1986). A third component—technological knowledge—must be added to this toolbox to keep up with the 21st century. (Chapter 7)

Transmission model of instruction: A theory of learning that suggests that students will learn facts, concepts, and understandings by absorbing the content of their teacher's explanations (direct instruction) or by reading explanations from a text and answering related questions. Teachers tell information to students in this model, and instruction is teacher led and teacher directed. (Chapter 2)

Tri-mind activities: Activities developed by Sternberg and based on his theory of triarchic theory of intelligence. Activities can be designed to cater to the analytical, creative, or practical thinker. Sternberg and Grigorenko (2000) state that students can learn more effectively if they are taught in a way that matches their thinking style. (Chapter 5)

Two-dimensional model of instruction: An educational model that focuses on the facts and content of the subject and the rote memorization of procedures and topics. A two-dimensional curriculum and instruction model focuses on the bottom levels of the Structure of Knowledge and the Structure of Process. This encourages students to work at a low-order level of thinking (e.g., memorization of facts or perfunctory performance of lower level skills) in a content/skill–based, coverage-centered curriculum. A two-dimensional model often presents the generalization or new concepts at the beginning of the learning cycle and follows a direct teaching methodology. (Chapters 1 and 2)

Unit web: An essential tool for planning a concept-based unit of work. A unit web contains critical content topics and concepts and gives an overview of the depth and breadth of the unit of instruction. (Chapter 4)

Teaching for Deep Understanding in Secondary Schools Book Study

Participants' Metacognition Log

Margie Pearse has devised a metacognition log—a powerful tool for self-reflection when reading any text. I include her template here to guide your thoughts and help you process the main ideas from each chapter in this book. The template emphasizes a constructivist approach to learning, encouraging synthesis and self-discovery.

There are four main areas for reflection in the metacognition log:

VIPs: What were the very important points that stood out for you?

Making connections: What connections can you make between the ideas in the book and your own practice, life, or experience?

Ideas for my classroom: How can you apply the ideas to your own classroom?

Questioning: What questions did you ask yourself as you were reading?

All the discussion questions from the end of each chapter are also included here for ease of use in a study group situation.

 You will find the concept-based mathematics metacognition log under the Book Study Resources section of the companion website.

A Participant's Metacognition Log

PARTICIPANT'S NAME: _____

CHAPTER TITLE _____

VIPs	Making Connections	Ideas for My Classroom	Questioning
Very important points & quotes that stood out for me. Include a page number for each VIP so your group can share along with you.	Connections I made either to my life, a lesson I taught, or something else I read or learned. Include a page number for each connection so your group can share in your connections.	What ideas are you ready to try? What ideas would you like to tweak to best fit your teaching or grade level? This is a great opportunity to share practical teaching ideas. Include page numbers so your book study members can follow along with you.	What questions came to mind when reading this section? Include page numbers & the question that came to mind when reading. If you found the answer, include what page helped you answer your question.

A SYMPHONY OF INSIGHT:

What insights and reflections did I gain from my book study members as a result of this chapter?

Template created by Margie Pearse, author of *Teaching Numeracy: 9 Critical Habits to Ignite Mathematical Thinking* (Corwin, 2011). Reprinted with permission.

A Participant's Metacognition Log

PARTICIPANT'S NAME: _____

CHAPTER TITLE _____

VIPs	Making Connections	Ideas for My Classroom	Questioning
Very important points & quotes that stood out for me. Include a page number for each VIP so your group can share along with you.	Connections I made either to my life, a lesson I taught, or something else I read or learned. Include a page number for each connection so your group can share in your connections.	What ideas are you ready to try? What ideas would you like to tweak to best fit your teaching or grade level? This is a great opportunity to share practical teaching ideas. Include page numbers so your book study members can follow along with you.	What questions came to mind when reading this section? Include page numbers & the question that came to mind when reading. If you found the answer, include what page helped you answer your question.

A SYMPHONY OF INSIGHT:

What insights and reflections did I gain from my book study members as a result of this chapter?

Template created by Margie Pearse, author of *Teaching Numeracy: 9 Critical Habits to Ignite Mathematical Thinking* (Corwin, 2011). Reprinted with permission.

A Participant's Metacognition Log

PARTICIPANT'S NAME: _____

CHAPTER TITLE _____

VIPs	Making Connections	Ideas for My Classroom	Questioning
Very important points & quotes that stood out for me. Include a page number for each VIP so your group can share along with you.	Connections I made either to my life, a lesson I taught, or something else I read or learned. Include a page number for each connection so your group can share in your connections.	What ideas are you ready to try? What ideas would you like to tweak to best fit your teaching or grade level? This is a great opportunity to share practical teaching ideas. Include page numbers so your book study members can follow along with you.	What questions came to mind when reading this section? Include page numbers & the question that came to mind when reading. If you found the answer, include what page helped you answer your question.

A SYMPHONY OF INSIGHT:

What insights and reflections did I gain from my book study members as a result of this chapter?

Template created by Margie Pearse, author of *Teaching Numeracy: 9 Critical Habits to Ignite Mathematical Thinking* (Corwin, 2011). Reprinted with permission.

A Participant's Metacognition Log

PARTICIPANT'S NAME: _____

CHAPTER TITLE _____

VIPs	Making Connections	Ideas for My Classroom	Questioning
Very important points & quotes that stood out for me. Include a page number for each VIP so your group can share along with you.	Connections I made either to my life, a lesson I taught, or something else I read or learned. Include a page number for each connection so your group can share in your connections.	What ideas are you ready to try? What ideas would you like to tweak to best fit your teaching or grade level? This is a great opportunity to share practical teaching ideas. Include page numbers so your book study members can follow along with you.	What questions came to mind when reading this section? Include page numbers & the question that came to mind when reading. If you found the answer, include what page helped you answer your question.

A SYMPHONY OF INSIGHT:

What insights and reflections did I gain from my book study members as a result of this chapter?

Template created by Margie Pearse, author of *Teaching Numeracy: 9 Critical Habits to Ignite Mathematical Thinking* (Corwin, 2011). Reprinted with permission.

A Participant's Metacognition Log

PARTICIPANT'S NAME: _____

CHAPTER TITLE _____

VIPs	Making Connections	Ideas for My Classroom	Questioning
Very important points & quotes that stood out for me. Include a page number for each VIP so your group can share along with you.	Connections I made either to my life, a lesson I taught, or something else I read or learned. Include a page number for each connection so your group can share in your connections.	What ideas are you ready to try? What ideas would you like to tweak to best fit your teaching or grade level? This is a great opportunity to share practical teaching ideas. Include page numbers so your book study members can follow along with you.	What questions came to mind when reading this section? Include page numbers & the question that came to mind when reading. If you found the answer, include what page helped you answer your question.

A SYMPHONY OF INSIGHT:

What insights and reflections did I gain from my book study members as a result of this chapter?

Template created by Margie Pearse, author of *Teaching Numeracy: 9 Critical Habits to Ignite Mathematical Thinking* (Corwin, 2011). Reprinted with permission.

A Participant's Metacognition Log

PARTICIPANT'S NAME: _____

CHAPTER TITLE _____

VIPs	Making Connections	Ideas for My Classroom	Questioning
Very important points & quotes that stood out for me. Include a page number for each VIP so your group can share along with you.	Connections I made either to my life, a lesson I taught, or something else I read or learned. Include a page number for each connection so your group can share in your connections.	What ideas are you ready to try? What ideas would you like to tweak to best fit your teaching or grade level? This is a great opportunity to share practical teaching ideas. Include page numbers so your book study members can follow along with you.	What questions came to mind when reading this section? Include page numbers & the question that came to mind when reading. If you found the answer, include what page helped you answer your question.

A SYMPHONY OF INSIGHT:

What insights and reflections did I gain from my book study members as a result of this chapter?

Template created by Margie Pearse, author of *Teaching Numeracy: 9 Critical Habits to Ignite Mathematical Thinking* (Corwin, 2011). Reprinted with permission.

A Participant's Metacognition Log

PARTICIPANT'S NAME: _____

CHAPTER TITLE _____

VIPs	Making Connections	Ideas for My Classroom	Questioning
Very important points & quotes that stood out for me. Include a page number for each VIP so your group can share along with you.	Connections I made either to my life, a lesson I taught, or something else I read or learned. Include a page number for each connection so your group can share in your connections.	What ideas are you ready to try? What ideas would you like to tweak to best fit your teaching or grade level? This is a great opportunity to share practical teaching ideas. Include page numbers so your book study members can follow along with you.	What questions came to mind when reading this section? Include page numbers & the question that came to mind when reading. If you found the answer, include what page helped you answer your question.

A SYMPHONY OF INSIGHT:

What insights and reflections did I gain from my book study members as a result of this chapter?

Template created by Margie Pearse, author of *Teaching Numeracy: 9 Critical Habits to Ignite Mathematical Thinking* (Corwin, 2011). Reprinted with permission.

A Participant's Metacognition Log

PARTICIPANT'S NAME: _____

CHAPTER TITLE _____

VIPs	Making Connections	Ideas for My Classroom	Questioning
Very important points & quotes that stood out for me. Include a page number for each VIP so your group can share along with you.	Connections I made either to my life, a lesson I taught, or something else I read or learned. Include a page number for each connection so your group can share in your connections.	What ideas are you ready to try? What ideas would you like to tweak to best fit your teaching or grade level? This is a great opportunity to share practical teaching ideas. Include page numbers so your book study members can follow along with you.	What questions came to mind when reading this section? Include page numbers & the question that came to mind when reading. If you found the answer, include what page helped you answer your question.

A SYMPHONY OF INSIGHT:

What insights and reflections did I gain from my book study members as a result of this chapter?

Template created by Margie Pearse, author of *Teaching Numeracy: 9 Critical Habits to Ignite Mathematical Thinking* (Corwin, 2011). Reprinted with permission.

Discussion Questions

Part 1. What Is Concept-Based Curriculum and Instruction in Mathematics? Research and Theory

Chapter 1. Why Is It Important for My Students to Learn Conceptually?

1. Why does math education need to undergo a reform? Why or Why not?

2. Why do educators need to include the conceptual understandings of a topic represented in a three-dimensional curriculum model?

3. How do the Structures of Knowledge and Process apply to the mathematics realm?

4. What are the features of inductive teaching and the benefits of an inductive approach when learning mathematics?

5. How does synergistic thinking develop intellect?

6. How would you use the different levels of inquiry in your classroom? Think of examples of when you might use each (structured, guided, open).

Chapter 2. What Are the Levels of the Structures of Knowledge and Process for Mathematics?

1. What are facts in mathematics?

2. What is the difference between formulae and generalizations in mathematics?

3. What is the distinction between processes, skills, and algorithms in mathematics?

4. What are the key categories of processes in mathematics? Provide examples for each process.

5. How do the Structure of Knowledge and the Structure of Process represent different facets of learning mathematics? Explain the symbiotic relationship between the two structures for learning mathematics.

6. How is mathematics a language of conceptual relationships made of macro, meso, and micro concepts?

7. Where on the Structure of Knowledge and the Structure of Process should curriculum and instruction focus? Why?

Part 2. How to Craft Generalizations and Plan Units of Work to Ensure Deep Conceptual Understanding

Chapter 3. What Are Generalizations in Mathematics?

1. What is the difference between a principle and a generalization in mathematics?

2. Why do we wish our students to understand principles and generalizations in mathematics?

3. What is the difference between level 1, 2, and 3 generalizations?

4. What opportunities do you provide for your students to demonstrate and communicate their conceptual understanding?

5. How will you develop your skills at crafting generalizations when planning your units of work?

Chapter 4. How Do I Plan Units of Work for a Concept-Based Curriculum?

1. How does unit planning empower teachers?

2. What are the benefits of a unit-webbing tool?

3. What are the main components for a math unit web?

4. Why do math unit webs include concepts in mathematical processes?

5. How do the three types of essential questions—factual, conceptual, and debatable—support student learning?

6. What is the most effective way to create a unit of work? Describe the process that you would utilize.

7. How will you encourage collaboration when unit planning?

Part 3. How Do We Engage Students Through Instructional Practice? Strategies to Engage and Assess

Chapter 5. How Do I Captivate Students? Eight Strategies for Engaging the Hearts and Minds of Students

1. What strategies do you employ to engage and motivate the hearts and minds of your students?

2. How do you keep students on task in a social learning environment?

3. How do you promote a growth mindset in your classroom and encourage mistakes as part of the learning process?

4. Why do you think it is important to balance whole teacher talk time with student working time?

5. What differentiation strategies do you use to cater to the individual needs of your students?

6. What formative assessment strategies do you employ in your classroom?

7. What role do you think the structure of your classroom plays in student learning?

Chapter 6. How Do I Know My Students Understand the Concepts? Assessment Strategies

1. Choose three of the assessment strategies listed in this chapter. When, why, and how would you use each in your own classroom practice?

2. What are some advantages of visible thinking routines?

3. How would you use a visible thinking routine?

4. Are there other graphic organizers you use that highlight student's conceptual understanding?

5. What math topics lend themselves well to performance assessment tasks?

6. How do you develop core skills such as communication, self-management, and research skills in the math classroom? What might you do in the future to improve your practice in developing these skills?

7. How might you use the rubrics in this chapter for developing the concept-based student in your own classroom context?

Chapter 7. How Do I Integrate Technology to Foster Conceptual Understanding?

1. How would you use the SAMR model in your own practice?

2. How would you apply the TPACK model when planning a unit of work?

3. There are so many digital tools on the market how do you decide which ones to use?

4. How do you integrate technology in the math classroom, in order to promote and enhance learning?

5. How do you ensure students still engage in discussions when using Google tools such as Google Docs or Google Slides?

6. When would you use technology in your practice and when would you not?

Chapter 8. What Do Ideal Concept-Based Mathematics Classrooms Look Like?

1. How will you foster a growth mindset in your classroom?

2. How will you use the rubric for developing as a concept-based teacher and developing concept-based instruction in your own practice?

3. What does ideal math learning look like in your classroom?

4. What does ideal math teaching look like in your classroom?

5. What is the role of a teacher? Is it to transmit information and knowledge? Or is it to inspire, guide, and facilitate the social process of learning?

6. How will you lead the transition towards concept-based math curriculum for teaching deeper understanding?

7. How will you prepare your students for an unknown future?

References
and Further Reading

Artigue, M., & Baptist, P. (2010). *Inquiry in mathematics education*. Retrieved from www.fibonacci-project.eu

Austin, K., Darling-Hammond, L., Orcutt, S., & Martin, D. (2003). *Learning from others: Learning in a social context*. Retrieved from http://www.learner.org/courses/learningclassroom/support/07_learn_context.pdf

Banchiand, H., & Bell, R. (2008). The many levels of inquiry. *Science and Children, 26–29*. Retrieved from http://www.miseagrant.umich.edu/lessons/files/2013/05/The-Many-Levels-of-Inquiry-NSTA-article.pdf

Barron, B., & Darling-Hammond, L. (2008). *Teaching for meaningful learning: A review of research on inquiry-based and cooperative learning*. San Francisco: Jossey-Bass. Retrieved from http://www.edutopia.org/pdfs/edutopia-teaching-for-meaningful-learning.pdf

Beere, J. (2013). *The perfect (Ofsted) lesson*. Carmarthen, United Kingdom: Independent Thinking Press.

Belk, J. [Jim.belk]. (2010). *Unit circle angles color* [Image file]. Retrieved from https://commons.wikimedia.org/wiki/File:Unit_circle_angles_color.svg#/media/File:Unit_circle_angles_color.svg

Berliner, D. C. (2004). Describing the behavior and documenting the accomplishments of expert teachers. *Bulletin of Science, Technology & Society, 24*(3), 200–212.

Blackwell, L. S., Dweck, C. S., & Trzesniewski, K. H. (2007). Implicit theories of intelligence predict achievement across an adolescent. *Child Development, 78*(1), 246–263.

Blair, A. (2008). Inquiry teaching. *Mathematics Teaching Incorporating Micromath, 211,* 8–11. Retrieved from http://www.inquirymaths.com/home/articles-and-blog-posts

Blair, A. (2015, February 23). *Inquiry maths*. Retrieved from http://www.inquirymaths.co.uk

Boaler, J. (1998). Open and closed mathematics: Student experiences and understandings. *Journal for Research in Mathematics Education, 29*(1), 41–62.

Boaler, J. (2010). *The elephant in the classroom: Helping children learn and love maths*. London: Souvenir Press.

Boaler, J. and Dweck, C. (2015). Mathematical Mindsets: Unleashing Students' Potential through Creative Math, Inspiring Messages and Innovative Teaching. San Francisco: CA: Jossey-Bass.

Borovik, A. V., & Gardiner, T. (2007). *Mathematical abilities and mathematical skills*. Retrieved from http://www.maths.manchester.ac.uk/~avb/pdf/abilities2007.pdf

Bransford, J. D., Brown, A. L., & Cocking, R. R. (Eds.). (2000). *How people learn: Brain, mind, experience, and school*. Washington, D.C.: National Academy Press. Retrieved from http://www.nap.edu/books/0309070368/html

Bray, L. (2007). *The sweet far thing.* New York: Delacorte Press.

Casner-Lotto, J., & Benner, M. W. (2006). *Are they really ready to work?* New York, NY: The Conference Board, Partnership for 21st Century Skills, Corporate Voice for Working Families, and the Society for Human Resources Management.

Celizic, M. (2012). Meet the girl with half a brain. *Today.* Retrieved from http://www.today .com/id/36032653/ns/today-today_health/t/meet-girl-half-brain/#.VXuPuGSqqko

Dreyfus, H., & Dreyfus, S. E. (2000). *Mind over machine.* New York: The Free Press.

Dweck, C. S. (2006). *Mindset: The new psychology of success.* New York: Random House.

Dweck, C. S. (2012). *Mindset: How you can fulfil your potential.* London: Robinson.

Erickson, H. L. (2002). *Concept-based curriculum and instruction: Teaching beyond the facts.* Thousand Oaks, CA: Corwin.

Erickson, H. L. (2007). *Concept-based curriculum and instruction for the thinking classroom.* Thousand Oaks, CA: Corwin.

Erickson, H. L. (2008). *Stirring the head, heart and souls: Redefining curriculum, instruction and concept-based learning* (3rd ed). Thousand Oaks, CA: Corwin.

Erickson, H. L., & Lanning, L. (2014). *Transitioning to a concept-based curriculum and instruction: How to bring content and process together.* Thousand Oaks, CA: Corwin.

Ferguson, K. (2010). *Inquiry-based mathematics instruction versus traditional mathematics instruction: The effect on student understanding and comprehension in an eighth grade pre-algebra classroom* (Master's thesis, Cedarville University, Cedarville, Ohio). Retrieved from http://digital-commons.cedarville.edu/education_theses/26

Frayer, D., Frederick, W., & Klausmeier, H. (1969). *A schema for testing the level of cognitive mastery.* Madison, WI: Wisconsin Center for Education Research.

Frey, C. B., & Osborthe, M. (2013). *Future of employment: How susceptible are jobs to computerisation?* Oxford Martin School working paper, Oxford University. Retrieved from http:// www.oxfordmartin.ox.ac.uk/downloads/academic/The_Future_of_Employment.pdf

Fullan, M. (2006). *Change theory: A force for school improvement.* Centre for Strategic Education, Seminar Series Paper No. 157.

Fullan, M. (2008). *The six secrets of change.* Retrieved from http://www.michaelfullan.ca/ images/handouts/2008SixSecretsofChangeKeynoteA4.pdf

Gao, X. F., & Bao, J. S. (2012). *Research on concept-based instruction of calculus.* Paper presented at the 12th International Congress on Mathematical Education, Korea.

Gelman, B. D., Gruber, M. J., & Ranganath, C. (2014). States of curiosity modulate hippocampus-dependent learning via the dopaminergic circuit. *Neuron, 84*(2), 486–496.

Good, C., Aronson, J., & Inzlicht, M. (2003). Improving adolescents' standardized test performance: An intervention to reduce the effects of stereotype threat. *Applied Developmental Psychology, 24,* 645–662.

Hamdan, N., McKnight, P., McKnight, K., & Arfstrom, P. (2013). *A review of flipped learning.* Retrieved from http://www.flippedlearning.org/review

Hart Research Associates. (2013). *It takes more than a major: Employer priorities for college learning and student success.* Retrieved from http://www.aacu.org/sites/default/files/files/ LEAP/2013_EmployerSurvey.pdf

Harvard University, Project Zero. (2007). *Visible thinking in action.* Retrieved from http://www .visiblethinkingpz.org/VisibleThinking_html_files/01_VisibleThinkingInAction/01a_ VTInAction.html

Hattie, J. A. C. (2009). Visible learning: A synthesis of 800+ meta analyses on achievement. London: Routledge.

Hattie, J. A. C. (2012). *Visible learning for teachers: Maximizing impact on learning*. London: Routledge.

Holston, V., & Santa, C. (1985). Raft: A method of writing across the curriculum that works. *Journal of Reading, 28,* 456–457.

The Hong Kong Education Bureau. (2001). *The five higher order thinking skills*. Retrieved from http://cd1.edb.hkedcity.net/cd/maths/en/ref_res/material/hots_e/5Skill_e.pdf

Hyde, A. (2006). *Comprehending math: Adapting reading strategies to teach mathematics, K–6*. Portsmouth, NH: Heinemann.

Innovation enters the classroom. (n.d.). *Business Furniture*. Retrieved from http://business furniture.net/uncategorized/innovation-enters-the-classroom/

International Baccalaureate. (2014). *Mathematics*. Retrieved from http://www.ibo.org/en/programmes/diploma-programme/curriculum/mathematics/mathematics

International Baccalaureate. (2015). *Approaches to teaching and learning*. Retrieved from http://www.ibo.org/globalassets/digital-tookit/flyers-and-artworks/approaches-to-teaching-learning-dp-en.pdf

Kalchman, M., Fuson, K. C., & Bransford, J. D. (2005). *How students learn: Mathematics in the classroom*. Washington, D.C.: National Academy Press. Retrieved from http://www.nap.edu/catalog.php

Keeley, P., & Tobey, C. R. (2011). *Mathematics formative assessment strategies: 75 practical strategies for linking assessment, instruction and learning*. Thousand Oaks, CA: Corwin.

Khan Academy. (2014, February 12). *Introduction to complex numbers* [Video file]. Retrieved from https://www.youtube.com/watch?v=SP-YJe7Vldo

King, L. (2014). *Teaching with ATL in mind*. Retrieved from http://www.taolearn.com/atl_resources.php

Koehler, M. J. (2012). *Using the TPACK image*. Retrieved from http://www.matt-koehler.com/tpack/using-the-tpack-image/

Koehler, M. J., Mishra, P., & Cain, W. (2013). What is technological pedagogical content knowledge (TPACK)? *Journal of Education, 193*(3), 13–19.

Koehler, M. J., & Mishra, P. (2009). What is technological pedagogical content knowledge? *Contemporary Issues in Technology and Teacher Education, 9*(1), 60–70.

Kyriacou, C., & Issitt, J. (2007). Teacher-pupil dialogue in mathematics lessons. *British Society for Research into Learning Mathematics, 27*(3), 61–65.

Lanning, L. (2013). *Designing a concept-based curriculum in English language arts: Meeting the common core with intellectual integrity, K–12*. Thousand Oaks, CA: Corwin.

Laud, L. (2011). *Using formative assessment to differentiate mathematics instruction*. Thousand Oaks, CA: Corwin.

Maslow, A. (1987). *Motivation and personality* (3rd ed.). Boston, MA: Addison-Wesley.

Mason, J., Drury, H., & Bills, L. (2007). *Studies in the zone of proximal awareness*. Retrieved from http://www.merga.net.au/documents/keynote42007.pdf

McWilliam, E. (2009). Teaching for creativity: From sage to guide to meddler. *Asia Pacific Journal of Education, 29*(3), 281–293.

Montague, M., & Van Garderen, D. (2003). Visual-spatial representation, mathematical problem solving and students of varying abilities. *Learning Disabilities Research & Practice, 18*(4), 246–254.

National Council of Teachers of Mathematics. (2014). *Principles to actions: Ensuring mathematical success for all*. Retrieved from http://www.nctm.org/principlestoactions/

Nebeuts, E. (2015). *Inspirational Quotes, Words, Sayings*. Retrieved from http://www.inspirational stories.com/quotes/e-kim-nebeuts-to-state-a-theorem-and-then-to/

Number talks toolkit. (2015). *Math Perspectives.* Retrieved from http://www.mathperspectives .com/num_talks.html

Nunley, K. (2004). *Layered curriculum: The practical solutions for teachers with more than one student in their classroom* (2nd ed.). Amherst, NH: Brains.org.

Nunley, K. (2006). *Differentiating the high school classroom.* Thousand Oaks, CA: Corwin.

Nunley, K. (2011). *Enhancing your layered curriculum classroom: Tips, tune ups and technology.* Amherst, NH: Brains.org.

Ollerton, M. (2013). *Enquiry-based learning.* Retrieved September 10, 2015, from http:// www.mikeollerton.com/pubs/EBL.pdf

Ollerton, M., & Vasile, D. (2014, January). Fluffy assessment, joined up thinking, vision and future aspiration. *Association of Teachers of Mathematics,* 11–13.

Parrish, S. (2014). *Number talks: Helping children build mental math and computation strategies, grades K–5, updated with common core connections.* Sausalito, CA: Maths Solutions.

Pink, D. (2005). *A whole new mind.* New York: Riverhead Books.

Pólya, G. (1957). *How to solve it.* Garden City, NY: Doubleday.

Prensky, M. (2002). *E-nough.* Retrieved from http://www.marcprensky.com/writing/ Prensky%20-%20e-Nough%20-%20OTH%2011-1%20March%202003.pdf

Prince, M. J., & Felder, R. M. (2006). Inductive teaching and learning methods: Definitions, comparisons, and research bases. *Journal of Engineering Education, 95*(2), 123–138.

Prince, M., & Felder, R. (2007). The many faces of inductive teaching and learning. *Journal of College Science Teaching, 36*(5), 14–20.

Puentedura, R. (2013). *SAMR: A contextualised introduction.* Retrieved from http://www .hippasus.com/rrpweblog/archives/2013/10/25/SAMRAContextualizedIntroduction.pdf

Pugalee, D. K. (2005). *Writing to develop mathematical understanding.* Norwood, MA: Christopher-Gordon.

Queensland University of Technology. (2014). *Libraries and learning spaces: Flexible collaborative learning.* Retrieved from https://www.qut.edu.au/campuses-and-facilities/gardens-point-campus/libraries-and-learning-spaces

Quigley, A. (2013). *Dirty work.* Retrieved from http://www.huntingenglish.com/2013/10/12/ dirty-work/

Reys, R., Lindquist, M., Lambdin, D., & Smith, N. (2014). *Helping children learn mathematics* (11th ed.). Hoboken, NJ: Wiley.

Ritchhart, R., Church, M., & Morrison, K. (2011). *Making thinking visible: How to promote engagement, understanding and independence for all learners.* San Francisco: Jossey-Bass.

Ritchhart, R., & Perkins, D. (2008). Making thinking visible. *Educational Leadership, 65*(5), 57–61.

Roanau, R., Meyer, D., & Crites, T. (2014). *Putting essential understanding of functions into practice in grades 9-12.* Reston, VA: National Council of Teachers of Mathematics.

Rooney, C. (2012). How am I using inquiry-based learning to improve my practice and to encourage higher order thinking among my students of mathematics? *Educational Journal of Living Theories, 5*(2), 99–127.

Rowling, J. K. (2008). *The fringe benefits of failure and the importance of imagination* [Speech]. Retrieved from http://www.jkrowling.com/en_GB/#/timeline/harvard-commencement-address/

Scott, C. (2009). Talking to learn: Dialogue in the classroom. *The Digest, 2,* 1–2. Retrieved from http://www.nswteachers.nsw.edu.au

Scusa, T. (2008). *Five processes of mathematical thinking.* Math in the Middle Institute Partnership Action Research Project Report. Lincoln: University of Nebraska.

Shulman, L. S. (1986). Those who understand: Knowledge growth in teaching. *Educational Researcher, 15*(2), 4–31.

Sousa, D. A. (2015). *How the brain learns mathematics.* Thousand Oaks, CA: Corwin.

Stephenson, J. (1998). *The concepts of capability and its importance in higher education.* Retrieved from http://www.hear.ac.uk/assets/Documents/resources/heca/heca_cq_01.pdf

Sternberg, R., & Grigorenko, E. L. (2000). *Teaching for successful intelligence.* Arlington Heights, VA: Skylight. Retrieved from https://castl.duq.edu/Conferences/Library03/PDF/Intelligence/Sternberg_R_Grigorenko_E.pdf

Stonewater, J. K. (2002, November). The mathematics writer's checklist: The development of a preliminary assessment tool for writing in mathematics. *School Science and Mathematics, 102,* 324–334.

Swan, M. (2005). *Improving learning in mathematics: Challenges and strategies.* London: Department for Education and Skills Standards Unit.

Swan, M. (2006a). *Collaborative learning in mathematics.* Nottingham, UK: University of Nottingham.

Swan, M. (2006b). *Collaborative learning in mathematics: A challenge to our beliefs and practices.* London: National Research and Development Centre for Adult Literacy and Numeracy.

Swan, M. (2010). *Research for teacher: Collaborative mathematics.* London: General Teaching Counsel for England.

Thurston, W. (1990). Mathematical education. *Notices of the AMS, 37*(7), 844–850. Retrieved from http://arxiv.org/pdf/math/0503081.pdf

Tobey, C. R., & Arline, C. B. (2014a). *Uncovering Student Thinking About Mathematics in the Common Core: 25 Assessment Probes, Grades 6–8.* Thousand Oaks, CA: Corwin.

Tobey, C. R., & Arline, C. B. (2014b). *Uncovering Student Thinking About Mathematics in the Common Core: 25 Assessment Probes, High School.* Thousand Oaks, CA: Corwin.

Tomlinson, C., & Allan, S. (2000). *Leadership for differentiating schools and classrooms.* Alexandria, VA: ASCD.

Tomlinson, C. A., & Imbeau, M. B. (2010). *Leading and managing a differentiated classroom.* Alexandria, VA: ASCD.

Treisman, U. (1992). Studying students studying calculus: A look at the lives of minority mathematics students in college. *The College Mathematics Journal, 23*(5), 362–372.

U.S. National Mathematics Advisory Panel. (2008). *Foundations for success: The final report of the National Mathematics Advisory Panel.* Washington, DC: U.S. Department of Education.

Vygotsky, L. S. (1978). *Mind in society: The development of higher psychological processes.* Cambridge, MA: Harvard University Press. Retrieved from http://www.psy.cmu.edu/~siegler/vygotsky78.pdf

Warren-Price, T. (2003). *Action research: Investigating the amount of teacher talk in my classroom.* Unpublished manuscript, MA TEFL/TESL Distance Learning Programme, University of Birmingham, Birmingham. UK. Retrieved from http://www.birmingham.ac.uk/documents/college-artslaw/cels/essays/languageteaching/warren1.pdf

Wiggins, G., & McTighe, J. (2006a). *Understanding by design: Expanded 2nd ed.* Upper Saddle River, NJ: Pearson Merrill Prentice Hall.

Wiggins, G., & McTighe, J. (2006b). *Understanding by design: Professional development workbook.* Australia: Hawker Brownlow Education.

Wiggins, G., & McTighe, J. (2013). *Essential questions: Opening doors to student understanding.* Alexandria, VA: Association for Supervision and Curriculum Development.

Willingham, D. T. (2010). Is it true some people just can't do math? *American Educator,* 14–39. Retrieved from http://www.aft.org//sites/default/files/periodicals/willingham.pdf

Index

A

Acceleration, 109 (figure)

Accountability, student, 149

Activity-oriented curriculum, 156

Affective skills, 177

"Agree, Disagree, and Depends" model, 172, 176, 176 (figure)

Algebraic tools, 64–65, 65 (figure)

Algorithms
  knowledge-process relationship, 53–54, 55 (figure)
  mathematical processes, 35 (figure), 35–36
  order of operations rule, 36 (figure)
  rote memorization, 30

Allan, S., 232

Analytical thinkers, 144, 146 (figure), 146–147

"Approaches to Teaching and Learning" program, 177–178, 179 (figure), 218

Apps (applications)
  concept maps, 203–204
  Edmodo, 204–205
  entrance and exit tickets, 204
  mini interactive electronic whiteboards, 204

Arfstrom, P., 192

Arizona State University (ASU), 152

Arline, C. B., 176, 176 (figure)

Aronson, J., 216

Artigue, M., 255

Assessment strategies
  "Agree, Disagree, and Depends" model, 172, 176, 176 (figure)
  challenges, 156–157
  concept attainment cards, 172, 175 (figure)
  concept-based lesson planning, 224 (figure), 226–227 (figure)
  concept-based students, 178, 180, 181–182 (figure)
  core transdisciplinary skills, 177–178, 179 (figure)
  Frayer Model, 172–173 (figure), 174
  inquiry-based assessments, 157–158, 159 (figure)
  learning curve tool, 149–150, 150 (figure)
  multiple choice questions, 177

  open inquiry tasks, 158, 160
  performance assessment tasks, 143, 164–166, 165 (figure), 166 (figure), 167–171 (figure)
  self-assessments, 180, 183–184 (figure), 185
  visible thinking routines, 147, 149, 160–164, 161 (figure), 162 (figure)
  "Zero, One, Two, or Three" model, 176–177

Austin, K., 255

Australia, 32, 33–34 (figure)

Authentic performance assessments, 143, 164–166, 165 (figure), 166 (figure), 167–171 (figure)

B

Backward design, 87

Banchi, H., 15

Bao, J. S., 2, 229

Baptist, P., 255

Barron, B., 14, 231

Beere, J., 185

Belk, J., 209 (figure)

Bell, R., 15

Bengey, Ian, 135

Benner, M. W., 177, 231

Berliner, D. C., 178, 179 (figure)

Big ideas, 62

Bills, L., 62

Binge, Chris, 9

Blackwell, L. S., 255

Blair, A., 15, 49

Boaler, J., 255, 256

BODMAS (brackets, order, divide, multiply, add, and subtract), 35, 36, 133

Borovik, A. V., 21

Brain plasticity, 133–134

Bransford, J. D., 7, 12, 22

Bray, L., 152

Briggs, Ron, 152

Brown, A. L., 12
  see also Bransford, J. D.

Building height exercise, 166, 167 (figure), 170–171 (figure)

Butler, Douglas, 192

C
Cain, W., 191
Calculus
    fundamental theorems, 28–29, 29 (figure),
        110 (figure)
    gradients and slopes, 111–115 (figure)
    increasing/decreasing functions, 115 (figure)
    integration, 126 (figure)
    product rule, 109 (figure), 117–119 (figure)
    real-life problems, 120–125 (figure)
    stationary points, 103 (figure), 109 (figure),
        116 (figure)
    unit planning guidelines, 97, 104–107
        (figure), 108
    unit webs, 97, 103 (figure)
    weekly planner, 109–110 (figure)
Cartesian plane exercise, 96 (figure)
Casner-Lotto, J., 177, 231
Celizic, M., 134
Checklists, 87–88 (figure)
Chun Yu Yiu, 21
Church, M., 162 (figure), 163
Circle geometry, 97, 98 (figure), 99–102 (figure)
Circle theorems, 196–198 (figure)
Circular functions, 205
Classroom arrangement flexibility, 151–153,
    153 (figure)
Clintondale High School (Michigan), 192
Cocking, R. R., 12
    see also Bransford, J. D.
Cognitive closure, 163–164
Cognitive skills, 177
Colaiacovo, Rob, 218–219
Collaboration tools, 202–203
Collaborative learning, 131, 133, 199
Collaborative planning, 83
Common Core State Standards (CCSS), 32,
    33–34 (figure)
Communication skills, 35 (figure), 37 (figure),
    38, 41, 42 (figure), 43 (figure), 43–45
Comparison studies, 33–34 (figure)
Complex numbers, 200–201 (figure)
Compression, 28
Concept attainment cards, 174, 175 (figure)
Concept-based curriculum
    assessment strategies, 147, 149–150, 156–166
    basic concepts, xvi
    benefits, 228–229, 231–232
    challenges, 228–232
    engagement strategies, 131–153
    goals and outcomes, 230
    guiding principles, 22
    ideal classrooms, 215–221
    importance, 2–3
    unit planning guidelines, 83, 87–88 (figure)

Concept-based lesson planning, 224 (figure),
    225–227 (figure)
Concept-based students, 178, 180, 181–182
    (figure)
Concept maps, 203–204
Concepts, 6, 6 (figure), 29, 52, 58
Conceptual Age, 2
Conceptual lens
    basic concepts, 56, 85
    calculus, 97, 103 (figure), 104 (figure)
    circle geometry, 98 (figure), 99 (figure)
    functions, 84 (figure), 85
    mathematical processes, 85 (figure)
    unit planning guidelines, 90 (figure)
    unit web template, 86 (figure)
Conceptual questions
    calculus, 106–107 (figure)
    characteristics, 89
    circle geometry, 100–102 (figure)
    functions, 91–93 (figure)
"Connect, Extend, Challenge" thinking routine,
    162 (figure), 163
Content coverage, 156–157
Content knowledge (CK), 188, 190 (figure)
Content pedagogical knowledge (CPK), 188,
    190 (figure)
Core transdisciplinary skills, 177–178,
    179 (figure)
Crafting quality generalizations, 69, 71 (figure),
    71–74, 73 (figure)
Creating mathematical representations process
    comparison studies, 33–34 (figure)
    functional role, 45, 46 (figure), 47–49
    grid method, 49, 50–51 (figure)
    process generalizations, 7–9, 8 (figure)
    process relationships, 35 (figure)
    skills and strategies, 37 (figure)
Creative thinkers, 144
Criterion-referenced assessments, 158, 160
Crites, T., 259
Cubic functions, 138–140 (figure)
Cullinan, Ann Marie, 232
Curiosity, 14
Cylindrical tin exercise, 120 (figure),
    121 (figure), 123 (figure)

D
Darling-Hammond, L., 14, 231, 255
Debatable questions
    calculus, 107 (figure)
    characteristics, 89, 89 (figure)
    functions, 93 (figure)
Decreasing functions, 109 (figure), 115 (figure)
Deductive teacher-led instruction, 9, 11, 12
    (figure), 21

Depth of understanding, 26, 52, 72, 178, 180, 181 (figure), 220
Derivatives, 103 (figure)
Diagnostic assessments, 149, 150, 157
Differential calculus, 97, 103 (figure), 110 (figure), 126 (figure)
Differentiation strategies, 141, 143–144, 146–147, 224 (figure), 226–227 (figure), 232
D.I.R.T. (Dedicated Improvement and Reflection Time), 184 (figure), 185
Discriminants, 67, 68 (figure), 70 (figure)
Displacement, 109 (figure)
Division, algebraic, 64, 65, 65 (figure)
Double Angle task, 39–40 (figure)
Dougherty, B., 259
Dreyfus, H., 178, 179 (figure)
Dreyfus, S. E., 178, 179 (figure)
Drury, H.
    see Bills, L.
Dweck, C. S., 133–134, 216, 255

E
Edmodo, 204–205
Edwards, John, 74
Effective teaching practices, 217–218
Einstein, Albert, 133
Enduring understandings, 62, 63, 64 (figure)
Engagement strategies
    assessment practices, 149–150, 150 (figure)
    classroom arrangement flexibility, 151–153, 153 (figure)
    differentiation strategies, 141, 143–144, 146–147
    feedback techniques, 151
    hint cards/hint jars, 147, 148 (figure)
    inductive inquiry opportunities, 134–135, 136–140 (figure), 142 (figure)
    open, secure environments, 133–134
    social learning environments, 131, 133
    student choice, 147, 149
    teacher talk time (TTT), 135, 141
    tri-mind activities, 143, 144, 145 (figure), 146–147
Enquiry-based learning (EBL), 15, 158
Entrance tickets, 150, 162, 204
Equations
    perfomance tasks, 165 (figure)
    process generalizations, 65 (figure)
    quadratic equations, 67–69, 68 (figure), 70 (figure), 71–72, 75–78 (figure), 81 (figure), 136–140 (figure), 142 (figure)
    student learning experience, 96 (figure)
    unit planning guidelines, 90–93 (figure), 109 (figure)
    unit webs, 84 (figure), 103 (figure)
    visible thinking routines, 160–161

Erickson, H. L., 4, 5, 5 (figure), 7, 10 (figure), 11, 13, 14, 15, 22, 23, 32, 52, 53, 54, 57 (figure), 63, 66 (figure), 68 (figure), 69, 70 (figure), 79, 82, 83, 85, 87, 88, 88 (figure), 164, 165 (figure), 166, 178, 186, 218, 221, 224 (figure), 225–227 (figure), 233, 234
Erickson, Lynn, xvi
Essential understandings, 62, 63
Exit tickets, 150, 162, 204
Exponential functions, 66, 67 (figure)

F
Facts, 5, 6 (figure)
Factual knowledge, 26–28, 55 (figure), 56
Factual questions
    calculus, 106–107 (figure)
    characteristics, 89
    circle geometry, 100–102 (figure)
    functions, 91–93 (figure)
Feedback, 149, 151, 185
Felder, R. M., 230
Ferguson, K., 256
Financial problem, 120 (figure), 122 (figure), 124 (figure)
Fixed mindsets, 134
Flexible fronts, 151–153, 153 (figure)
Flipped classrooms, 192, 199, 200–201 (figure)
"Fluffy" assessment strategies, 157–158
Formative assessments, 149, 150
Formulae, 6, 27, 28
    see also Equations
Frayer, D., 174
Frayer Model, 172–173 (figure), 174
Frederick, W., 174
Frenis, Jan, 83
Frey, C. B., 3
Fullan, M., 150
Functions
    analytical thinking approach, 146 (figure), 146–147
    circular functions, 205
    mathematics graphing software, 192
    product rule, 109 (figure), 117–119 (figure)
    Structure of Knowledge, 6 (figure), 10 (figure), 56, 58
    Structure of Process, 8 (figure), 10 (figure), 56, 58 (figure)
    unit planning guidelines, 89, 90–93 (figure)
    unit webs, 84 (figure), 85
    weekly planner, 94–95 (figure), 109 (figure)
Fundamental theorems, 28–29, 29 (figure), 110 (figure)
Fuson, K. C., 7
    see also Bransford, J. D.

G

Gandhi, Mahatma, 215

Gao, X. F., 2, 229

Gardiner, T., 21

Gelman, B. D., 14, 219

Generalizations

    basic concepts, 6 (figure), 6–7, 8, 8 (figure),
        63–64

    benefits, 58

    calculus, 106–107 (figure)

    circle geometry, 100–102 (figure)

    crafting quality generalizations, 69, 71
        (figure), 71–74, 73 (figure)

    functional role, 62–63

    graphic organizers, 79, 80 (figure)

    guiding questions, 88–89, 89 (figure), 91–93
        (figure)

    inductive inquiry instruction, 74, 75–78
        (figure), 80 (figure), 81 (figure)

    mathematical examples, 64–69, 65 (figure),
        66 (figure), 67 (figure), 68 (figure), 70
        (figure)

    Structure of Knowledge, 26 (figure), 29–30

    three-dimensional inductive instruction
        models, 11–13, 12 (figure), 13 (figure)

    unit planning guidelines, 83, 90–93 (figure)

"Generate, Sort, Connect, and Elaborate"
    thinking routine, 163

GeoGebra, 192

Good, C., 216

Google applications, 202–203

Google Docs, 202–203

Google Forms, 203

Google Sheets, 203

Google Slides, 202

Gradients

    calculus weekly planner, 109 (figure)

    guiding questions, 89

    student learning experience, 111–116 (figure)

    unit planning guidelines, 90 (figure), 95
        (figure), 104–106 (figure)

    unit webs, 103 (figure)

Graphical display calculators, 191–192

Graphic organizers, 45, 46 (figure), 79, 80 (figure)

    see also Frayer Model

Graphing software

    benefits, 191–192

    circle theorems, 196–198 (figure)

    transformation of curves, 193–195 (figure)

GRASPS (goal, role, audience, situation,
    performance, standards) model, 166

Grid method, 49, 50–51 (figure)

Grigorenko, E. L., 143, 144

Growth mindsets, 134, 216, 231

Gruber, M. J., 14

    see also Gelman, B. D.

Guided inquiry tasks, 15, 16, 16 (figure), 17
    (figure), 19 (figure), 135, 141 (figure), 143

Guiding questions

    calculus, 97, 106–107 (figure)

    characteristics, 88–89, 89 (figure)

    circle geometry, 100–102 (figure)

    concept-based lesson planning, 224 (figure),
        225 (figure)

    functions, 91–93 (figure)

H

Hamdan, N., 192

Hart Research Associates, 3, 228

Harvard University, Project Zero, 147, 151, 160,
    163, 185

Hattie, J. A. C., 141

Headline routine, 163

High Concept level, 58

Higher-order thinking skills, 25, 33–34
    (figure), 35

Hint cards/hint jars, 147, 148 (figure)

Holston, V., 165, 166 (figure)

Hong Kong, 32, 33–34 (figure), 35

Hong Kong Education Bureau, 257

Hyde, A., 147

I

Ideal concept-based math lesson

    learning targets, 220–221, 222–223 (figure),
        224 (figure), 225 (figure)

    lesson closing, 221, 223 (figure), 224 (figure),
        225 (figure)

    lesson opening, 218–220, 222 (figure), 224
        (figure), 225 (figure)

Ideal math classrooms

    characteristics, 215–216

    effective teaching practices, 217–218

    growth mindset training, 216

    ideal math lesson, 218–220

Imaginary numbers, 200–201 (figure)

Imbeau, M. B., 143

Increasing functions, 109 (figure), 115 (figure)

Inductive inquiry instruction

    engagement strategies, 134–135, 136–140
        (figure), 142 (figure)

    generalizations, 74, 75–78 (figure), 80 (figure),
        81 (figure)

    mathematics graphing software, 192

Inductive instruction model, 11–12, 12
    (figure), 21

"Innovation Enters the Classroom", 151

Inquiry-based assessments, 157–158, 159
    (figure)

Inquiry-based learning

    basic concepts, 13–15, 14 (figure)

    benefits, 21

engagement strategies, 134–135, 136–140 (figure), 142 (figure)

generalizations, 74, 75–78 (figure), 80 (figure), 81 (figure)

ideal classrooms, 215–221

levels of inquiry, 15–17, 16 (figure), 17 (figure), 231–232

mathematical processes, 35 (figure)

Pythagorean theorem task, 18 (figure), 19 (figure), 20 (figure)

Inquiry levels
see Inquiry-based learning

Integral calculus, 97, 103 (figure), 110 (figure), 126 (figure)

International Baccalaureate, 32, 33–34 (figure), 35, 54 (figure), 85, 89, 158, 177–178, 186, 218, 233

Interpretations, 7, 8 (figure)

Investigating processes, 35, 35 (figure), 37 (figure), 49

Inzlicht, M., 216

iPads, 204

Issitt, J., 257

Iterative stages, 49, 52

"I used to think . . . Now I think . . . :" routine, 163

K

Kalchman, M., 7
see also Bransford, J. D.

Keeley, P., 174

Khan Academy, 199

Kinematics, 103 (figure), 109 (figure), 120 (figure), 122 (figure), 125 (figure)

King, L., 178, 179 (figure)

Klausmeier, H., 174

Knowledge-process relationship, 53–54, 55 (figure), 56, 57 (figure)

Know-understand-do (KUD) strategy
calculus, 97, 104–105 (figure)
circle geometry, 97, 99–100 (figure)
functions, 90–91 (figure), 94
inquiry process continuum, 14 (figure)

Koehler, M. J., 188, 191

K-W-L (know-want-learn) instructional strategy, 147, 149 (figure)

Kyriacou, C., 257

L

Lambdin, D., 259

Lanning, L., 4, 5 (figure), 7, 10 (figure), 11, 13, 15, 32, 52, 53, 57 (figure), 65 (figure), 67 (figure), 70 (figure), 79, 85, 87, 88 (figure), 164, 178, 181–182 (figure), 186, 218, 221, 222–223 (figure), 224 (figure), 225–227 (figure), 233, 234

Lanning, Lois, xvi–xvii

Laud, L., 157

Layered curriculum approach, 147, 149

Learning curve tool, 149–150, 150 (figure)

Learning space design, 151–153, 153 (figure)

Lesson plans
learning targets, 220–221, 222–223 (figure), 224 (figure), 225 (figure)
lesson closing, 221, 223 (figure), 224 (figure), 225 (figure)
lesson opening, 218–220, 222 (figure), 224 (figure), 225 (figure)

Levels of inquiry
see Inquiry-based learning

Lindquist, M., 259

Linear functions
basic concepts, 6
perfomance tasks, 165 (figure)
performance assessment tasks, 165 (figure), 166, 166 (figure)
student learning experience, 96 (figure)
unit planning guidelines, 90–93 (figure), 95 (figure)
unit webs, 84 (figure), 85, 103 (figure)

Logarithmic laws, 66–67, 67 (figure), 73 (figure)

M

Macro concepts, 52, 53 (figure), 55 (figure), 97

Making connections process, 35 (figure), 37 (figure), 45

Martin, D., 255

Maslow, A., 178, 179 (figure)

Mason, J.
see Bills, L.

Mastery level assessment, 178, 179 (figure), 180, 181–182 (figure), 222–223 (figure), 224 (figure), 225–227 (figure)

Mathematical processes
basic concepts, 32
calculus, 103 (figure)
circle geometry, 98 (figure)
comparison studies, 33–34 (figure)
creating representations, 7–9, 8 (figure)
six basic processes, 35 (figure), 35–36, 37 (figure), 38–39, 41, 43–45, 47–49
trigonometry, 31 (figure)
unit planning guidelines, 87 (figure), 97
unit webs, 84 (figure), 85

Mathematics graphing software
benefits, 191–192
circle theorems, 196–198 (figure)
transformation of curves, 193–195 (figure)

McKnight, K., 192

McKnight, P, 192

McTighe, J., 63–64, 87, 89, 156, 164, 165

McWilliam, E., 258

Memorization, 26–27, 30, 157

Meso concepts
calculus, 97, 103 (figure)

characteristics and functional role, 52, 53
(figure), 55 (figure)
circle geometry, 97
functions, 89
quadratics, 68
trigonometry, 57 (figure)
vectors, 65–66, 66 (figure)
Metacognition log, 243, 244–251 (figure)
Metacognitive learning approach, 22
Metacognitive skills, 177
Meyer, D., 259
Micro concepts
characteristics and functional role, 52, 53
(figure), 55 (figure)
functions, 58
International Baccalaureate diploma
mathematics courses, 54 (figure)
quadratics, 68–69
trigonometry, 80 (figure)
unit web template, 86 (figure)
weekly planner, 94–95 (figure)
Mind42, 191, 204
Mind maps, 203–204
Mini interactive electronic whiteboards, 204
Minority groups, 44
Mishra, P., 188, 191
Mistakes, 133–134
Mobile device apps
concept maps, 203–204
Edmodo, 204–205
entrance and exit tickets, 204
mini interactive electronic whiteboards, 204
Models
concept-based instruction, 222 (figure)
process generalizations, 7, 8 (figure)
unit planning guidelines, 90–92 (figure),
95 (figure)
unit webs, 84 (figure), 85
Montague, M., 48
Mora, Linda, 22
Morrison, K., 162–163
see also Church, M.
Mott, Cameron, 134
Multimedia projects, 199
Multiple choice questions, 177
Multiplication, algebraic, 64, 65, 65 (figure)

N
National Council of Teachers of Mathematics,
47, 62, 94, 97, 127, 133, 216, 217–218, 228,
231, 232, 233
Nebeuts, E., 21
Neuroplasticity, 133–134
Newmann, F., 231
"No no" verbs, 69, 71–72, 74
Number talks, 38, 41, 42 (figure),
43 (figure), 45

Number Talks Tool Kit, 38
Nunley, K., 147, 149

O
Ollerton, M., 15, 157–158
Open box problem, 120 (figure), 121 (figure),
123 (figure)
Open-ended questions, 160
Open inquiry tasks, 15, 16 (figure), 17, 17
(figure), 20 (figure), 135, 158, 160
Open, secure environments, 133–134
Orcutt, S., 255
Order of operations rule, 35–36, 36 (figure)
Osborthe, M., 3
Overarching generalizations, 64, 64 (figure)

P
Paddle pop stick technique, 151
Padlet, 191, 204
Parker, R., 38
Parrish, S., 258
Pearse, M., 243, 244–251 (figure)
Pedagogical content knowledge (PCK), 190
(figure), 191
Pedagogical knowledge (PK), 188, 190 (figure)
PEMDAS (parentheses, exponent, multiplication,
division, addition, and subtraction)
algorithm, 35–36, 36 (figure), 133
Performance assessment tasks, 143, 164–166,
165 (figure), 166 (figure), 167–171 (figure)
Perkins, D., 258
Personal exploration, 158, 160
Pictorial representations, 48–49
Pink, D., 2, 58
Placemat activity, 131, 132 (figure)
Plutarch, 157
Pólya, G., 36
Polygons, 173 (figure)
Polynomial functions, 135, 140 (figure), 141 (figure)
PPP (presentation, practice, and production)
teaching model, 9, 11, 12 (figure)
Practical thinkers, 144
Pre-assessments, 149, 150, 157
Prensky, M., 187
Prince, M. J., 230
Principles
basic concepts, 4, 5 (figure), 6 (figure), 6–7, 64
benefits, 58
concept-based instruction, 22
core transdisciplinary skills, 177–178
growth mindsets, 216
knowledge acquisition, 43–44
pedagogical principles, 217–218
problem solving, 36
theorems, 28, 29 (figure), 29–30
see also Generalizations; Structure of
Knowledge; Structure of Process

Problem-based learning methods, 12
Problem solving, 32, 33–34 (figure), 35 (figure),
    36, 37 (figure), 38
Procedural fluency, 9
Process-driven disciplines, 4
Processes
    basic concepts, 7
    knowledge-process relationship, 53–54, 55
        (figure), 56
Process generalizations
    crafting quality generalizations, 71–73, 73
        (figure)
    creating representations, 7–9, 8 (figure)
    equations, 65 (figure)
    logarithmic laws, 66, 67 (figure)
    problem solving, 65
    quadratic equations, 67–69, 68 (figure)
    unit planning guidelines, 83
    vectors, 65–66, 66 (figure)
Product rules, 109 (figure), 117–119 (figure)
Project Zero (Harvard University), 147, 151,
    160, 163, 185
Proofs, 35 (figure), 37 (figure), 38, 39–40 (figure)
Provocative questions, 89, 89 (figure)
Puentedura, R., 188, 189 (figure)
Pugalee, D. K., 79
Pythagorean theorem task
    factual knowledge, 27–28
    guided inquiry, 19 (figure)
    levels of inquiry, 17, 17 (figure)
    open inquiry, 20 (figure)
    order of operations rule, 36 (figure)
    structured inquiry, 18 (figure)

Q
Quadratic formulae/equations, 67–69, 68
    (figure), 70 (figure), 71–72, 75–78 (figure),
    81 (figure), 136–140 (figure), 142 (figure)
Queensland University of Technology (QUT), 152
Question Sorts, 147, 149
Quigley, A., 258

R
RAFTS (role, audience, format, topic, strong)
    model, 165–166, 166 (figure)
Ranganath, C., 14
    see also Gelman, B. D.
Rational numbers, 175 (figure)
Ratios, 62, 168–169 (figure)
Real-life problems, 120–125 (figure), 166,
    167–171 (figure)
Reasoning skills, 33–34 (figure), 35 (figure), 37
    (figure), 38, 39–40 (figure)
Representations
    see Creating mathematical representations
        process

Revisions, 7, 8 (figure)
Reys, R., 259
Richardson, K., 38
Right-angled trigonometry, 25, 26 (figure), 27,
    30, 80 (figure), 163, 168–169 (figure)
Ritchhart, R., 163, 258
    see also Church, M.
Ronau, R., 259
Rooney, C., 259
Roots of functions, 67–69, 68 (figure), 70 (figure),
    75–78 (figure), 136–141 (figure), 142 (figure)
Rote memorization, 26–27, 30, 157
Rowling, J. K., 21
Rubrics
    "Approaches to Teaching and Learning"
        assessment, 179 (figure)
    concept-based instruction, 222–223 (figure)
    concept-based lesson planning, 224 (figure),
        225–227 (figure)
    concept-based students, 181–182 (figure)

S
SAMR (substitution, augmentation,
    modification, and redefinition) model, 188,
    189 (figure)
Sanda, David, 3
Santa, C., 165, 166 (figure)
Scaffolding
    crafting quality generalizations, 69, 71
        (figure), 71–73, 73 (figure)
    structured inquiry, 143
Scalar products, 65–66, 66 (figure)
Schematic representations, 48–49
Schwartz, D., 30–31
Scott, C., 259
Scusa, T., 259
Secure environment, 133–134
Self-assessments, 180, 183–184 (figure), 185
Sequences and series, 73 (figure)
Shulman, L. S., 188
Sine curves, 206–212 (figure)
Six mathematical processes
    see Mathematical processes
Skills
    basic concepts, 7, 32
    knowledge-process relationship, 53–54, 55
        (figure)
    mathematical processes, 35, 35 (figure), 37 (figure)
Slopes
    calculus weekly planner, 109 (figure)
    guiding questions, 89
    student learning experience, 111–116
        (figure)
    unit planning guidelines, 90 (figure), 95
        (figure), 104–106 (figure)
    unit webs, 103 (figure)

Smith, N., 259
Social constructivist theory, 41, 43–44
Social learning environments, 131, 133
Socrates, 157
SOHCAHTOA, 25, 30, 163
Sousa, D. A., 44, 79, 141, 163, 164
Spaghetti activity, 205, 206–208 (figure), 213
    (figure)
Spherical balloon exercise, 120 (figure), 121
    (figure), 124 (figure)
Square roots, 67–69, 68 (figure), 70 (figure),
    75–78 (figure)
Statements of inquiry, 63
Statements Of Learning, 32, 33–34 (figure)
Stationary points, 103 (figure), 109 (figure), 116
    (figure)
Stephenson, J., 178, 179 (figure)
Sternberg, R., 143, 144
Stonewater, J. K., 79
Strategies
    basic concepts, 7, 31
    knowledge-process relationship, 53–54,
        55 (figure)
    mathematical processes, 35, 35 (figure),
        37 (figure)
    process generalizations, 81 (figure)
    see also Assessment strategies; Engagement
        strategies; Structure of Knowledge;
        Structure of Process
Structured inquiry tasks, 15, 16, 16 (figure), 17
    (figure), 18 (figure), 135, 136–140 (figure),
    142 (figure), 143
Structure of Knowledge
    basic concepts, 4–7, 5 (figure), 6 (figure)
    concept level, 29
    factual level, 26–28, 56
    functions, 6 (figure), 10 (figure), 56, 58
    generalizations, 26 (figure), 29–30
    knowledge-process relationship, 53–54, 55
        (figure), 56, 57 (figure)
    principles, 29 (figure), 29–30, 64
    quadratics, 68, 68 (figure), 70 (figure)
    topic level, 29, 53
    trigonometry, 26 (figure), 57 (figure)
    vectors, 65–66, 66 (figure)
Structure of Process
    basic concepts, 4, 5 (figure), 7–9, 8 (figure)
    equations, 65 (figure)
    functions, 8 (figure), 10 (figure), 56, 58
        (figure)
    knowledge-process relationship, 53–54, 55
        (figure), 56, 57 (figure)
    logarithmic laws, 66–67, 67 (figure), 73
        (figure)
    mathematical processes, 31 (figure), 32, 33–34
        (figure), 35 (figure), 35–36, 38, 41

process generalizations, 8, 8 (figure), 65
    (figure), 65–69, 67 (figure), 71–73, 73
    (figure), 83
quadratics, 68, 70 (figure)
skills, 32, 35, 35 (figure), 37 (figure)
strategies, 31, 35, 35 (figure), 37 (figure)
trigonometry, 31 (figure), 57 (figure)
Student accountability, 149
Student-centered instruction, 12, 12 (figure),
    21–22, 131, 133, 141
Student choice, 147, 149
Student engagement
    see Engagement strategies
Student learning experience
    circle theorems, 196–198 (figure)
    coordinates game, 96 (figure)
    Double Angle task, 39–40 (figure)
    functions, 115 (figure), 145 (figure), 148 (figure)
    gradients and slopes, 111–114 (figure)
    graphic organizers, 80 (figure)
    grid method, 50–51 (figure)
    guided inquiry task, 141 (figure)
    hint cards/hint jars, 148 (figure)
    imaginary and complex numbers, 200–201
        (figure)
    increasing/decreasing functions, 115 (figure)
    inductive inquiry, 75–78 (figure)
    integration, 126 (figure)
    measuring building height, 166, 167 (figure),
        170–171 (figure)
    number system, 46 (figure)
    number talks, 42–43 (figure)
    placement activity, 132 (figure)
    product rule, 117–119 (figure)
    rational numbers, 175 (figure)
    real-life problems, 120–125 (figure)
    self-assessment worksheet, 183–184 (figure)
    similar right-angled triangles, 168–169 (figure)
    spaghetti activity, 206–208 (figure), 213 (figure)
    stationary points, 116 (figure)
    structured inquiry task, 136–140 (figure)
    transformation of curves, 193–195 (figure)
    triangle inequality, 213 (figure)
    tri-mind activities, 145 (figure)
    unit circles, 206–208 (figure), 209–212 (figure)
    weekly planner, 94–95 (figure)
Substitutions, 7, 8 (figure)
Subtraction, algebraic, 64, 65, 65 (figure)
Summative assessments, 150
Swan, M., 131, 133
Synergistic thinking
    assessment strategies, 178, 180, 181 (figure)
    concept-based lesson planning, 224 (figure),
        225 (figure), 229, 234
    conceptual lenses, 56, 85, 87 (figure)
    inquiry process continuum, 14–15, 54

intellectual development, xvi, 14, 229, 232
learning targets, 220, 222–223 (figure)

T
Tablets, 204
Task commitment, 178, 180, 182 (figure)
Teacher talk time (TTT), 135, 141
Technological content knowledge (TCK), 190
    (figure), 190–191
Technological knowledge (TK), 188, 190 (figure)
Technological pedagogical knowledge (TPK),
    190, 190 (figure)
Technology integration
    benefits, 205
    flipped classrooms, 192, 199, 200–201 (figure)
    graphical display calculators, 191–192
    mathematics graphing software, 191–192,
        193–198 (figure)
    mobile device apps, 203–205
    multimedia projects, 199
    SAMR (substitution, augmentation,
        modification, and redefinition) model,
        188, 189 (figure)
    TPACK model, 188, 190 (figure), 190–191
Theorems, 7, 28–29, 29 (figure), 64, 110 (figure)
Theory, basic concepts, 4, 5 (figure)
Thinking styles, 144
"Think, Pair, Share" thinking routine, 151, 163
Three-dimensional curriculum and instruction
        model, 11–13, 12 (figure), 13 (figure), 58
"Thumbs-at-chest" technique, 151
Thurston, W., 28
Time-filling activities, 156, 164
Tobey, C. R., 174, 176, 176 (figure)
Tomlinson, C. A., 143, 232
Topical generalizations, 64, 64 (figure)
Topics, 4, 29, 53
TPACK model, 188, 190 (figure), 190–191
Transformation of curves, 193–195 (figure)
Transmission model of instruction, 56
Treisman, U., 44
Triangles
    Double Angle task, 39 (figure)
    right-angled triangles, 25, 26 (figure), 27, 30,
        31 (figure), 57 (figure), 80 (figure), 131,
        163, 168–169 (figure)
    triangle inequality, 213 (figure)
Trigonometry
    graphic organizers, 79, 80 (figure)
    measuring building height, 166, 167 (figure),
        170–171 (figure)
    performance assessment tasks, 166, 167 (figure)
    ratios, 62, 168–169 (figure)
    real-life problems, 166, 167–171 (figure)
    Structure of Knowledge, 26 (figure), 57 (figure)
    Structure of Process, 31 (figure), 57 (figure)
    see also Right-angled trigonometry

Tri-mind activities, 143, 144, 145 (figure), 146–147
Trzesniewski, K. H., 255
Two-dimensional curriculum and instruction
        model, 11, 12 (figure), 13, 13 (figure), 56, 58

U
Understanding By Design (UbD), 63, 87, 164
Unit circles, 206–208 (figure), 209–212 (figure)
United States, 32, 33–34 (figure)
Unit planning
    calculus, 97, 104–107 (figure), 108
    circle geometry, 97, 99–102 (figure)
    collaborative planning, 83, 89
    framework, 87–88
    functions, 89, 90–93 (figure)
    guiding questions, 88–89, 89 (figure), 91–93
        (figure)
    step-by-step checklist, 87–88 (figure)
    weekly planner, 94–95 (figure)
Unit webs
    calculus, 97, 103 (figure)
    circle geometry, 97, 98 (figure)
    conceptual lenses, 85, 86 (figure)
    functional role, 79, 83
    functions, 84 (figure), 85
U.S. National Mathematics Advisory Panel, 3

V
Van Garderen, D., 48
Vasile, D., 48, 157–158
Vectors, 65–66, 66 (figure)
Velocity, 109 (figure)
Venn diagrams, 45, 46 (figure), 190, 190 (figure)
Visible thinking routines, 147, 149, 160–164,
        161 (figure), 162 (figure)
Vygotsky, L. S., 41, 43

W
Warren-Price, T., 141
Waterbury School System (Connecticut), 30–31
Weekly planner
    calculus, 109–110 (figure)
    functions, 94–95 (figure)
"What, Why, How" model, 165 (figure), 166
Wiggins, G., 63–64, 87, 89, 156, 164, 166
Willingham, D. T., 5
Writing mathematics process, 44–45, 74, 79

X
X-intercepts, 68 (figure), 69, 70 (figure)

Y
YouTube, 199

Z
"Zero, One, Two, or Three" model, 176–177
Zeros of functions, 67–69, 68 (figure), 70 (figure)

A SAGE Company

CORWIN HAS ONE MISSION: to enhance education through intentional professional learning.

We build long-term relationships with our authors, educators, clients, and associations who partner with us to develop and continuously improve the best evidence-based practices that establish and support lifelong learning.

Solutions you want. Experts you trust. Results you need.